T0136993

Smart Innovation, Systems and Technologies

Volume 149

Series Editors

Robert J. Howlett, Bournemouth University and KES International,
Shoreham-by-sea, UK
Lakhmi C. Jain, Faculty of Engineering and Information Technology,
Centre for Artificial Intelligence, University of Technology Sydney,
Broadway, NSW, Australia

The Smart Innovation, Systems and Technologies book series encompasses the topics of knowledge, intelligence, innovation and sustainability. The aim of the series is to make available a platform for the publication of books on all aspects of single and multi-disciplinary research on these themes in order to make the latest results available in a readily-accessible form. Volumes on interdisciplinary research combining two or more of these areas is particularly sought.

The series covers systems and paradigms that employ knowledge and intelligence in a broad sense. Its scope is systems having embedded knowledge and intelligence, which may be applied to the solution of world problems in industry, the environment and the community. It also focusses on the knowledge-transfer methodologies and innovation strategies employed to make this happen effectively. The combination of intelligent systems tools and a broad range of applications introduces a need for a synergy of disciplines from science, technology, business and the humanities. The series will include conference proceedings, edited collections, monographs, handbooks, reference books, and other relevant types of book in areas of science and technology where smart systems and technologies can offer innovative solutions.

High quality content is an essential feature for all book proposals accepted for the series. It is expected that editors of all accepted volumes will ensure that contributions are subjected to an appropriate level of reviewing process and adhere to KES quality principles.

** Indexing: The books of this series are submitted to ISI Proceedings, EI-Compendex, SCOPUS, Google Scholar and Springerlink **

More information about this series at http://www.springer.com/series/8767

Xiaobo Qu · Lu Zhen · Robert J. Howlett ·
Lakhmi C. Jain
Editors

Smart Transportation Systems 2019

Springer

Editors
Xiaobo Qu
Department of Architecture
and Civil Engineering
Chalmers University of Technology
Gothenburg, Västra Götalands Län, Sweden

Robert J. Howlett
KES International Research
Bournemouth University
Poole, Dorset, UK

Lu Zhen
School of Management
Shanghai University
Shanghai, China

Lakhmi C. Jain
University of Canberra
Canberra, Australia

ISSN 2190-3018   ISSN 2190-3026   (electronic)
Smart Innovation, Systems and Technologies
ISBN 978-981-13-8685-5   ISBN 978-981-13-8683-1   (eBook)
https://doi.org/10.1007/978-981-13-8683-1

This Springer imprint is published by the registered company Springer Nature Singapore Pte Ltd.
The registered company address is: 152 Beach Road, #21-01/04 Gateway East, Singapore 189721, Singapore

Organisation

Honorary Chair

Lakhmi C. Jain, University of Canberra, Australia, and University of Technology Sydney, Australia

General Chair

X. Qu, Chalmers University of Technology, Sweden

Executive Chair

R. J. Howlett, Bournemouth University, UK

International Programme Committee

Dr. Ahmad Taher Azar, Faculty of Computers and Information, Benha University, Egypt
Assoc. Prof. Yiming Bie, Jilin University, China
Assist. Prof. Jyotir Moy Chatterjee, Asia Pacific University of Technology & Innovation, Nepal
Prof. Said Easa, Ryerson University, Canada
Dr. MM Haque, Queensland University of Technology, Australia
Dr. Yan Kuang, Lecturer in School of Engineering and Built Environment, Griffith University, Australia

Preface

With the newly developed emerging technologies, our transport and logistics sectors are undergoing revolutionary change, with intelligence, connectivity, automation, environment-friendly practice, etc. The 2nd International Symposium for Smart Transportation Systems provides a platform for researchers and practitioners to gather together and discuss the new revolutions. The topics range from macroscopic land use planning to microscopic traffic/intersection operations, from traditional modelling of traffic flow dynamics via car following models to the practice/applications of machine learning in transport systems, from logistics/routing to supply chain management, from vulnerable road users to heavy-duty vehicles, and from traditional transport network modelling to blockchain management in transport networks. The participants are mainly scholars and practitioners from Europe, China, Australia, and the USA.

Gothenburg, Sweden
Shanghai, China
Poole, UK
Canberra, Australia

Xiaobo Qu
Lu Zhen
Robert J. Howlett
Lakhmi C. Jain

Contents

About the Editors

Xiaobo Qu is a Professor and research group leader at the Division of Geology and Geotechnics, research group Urban Mobility Systems, Chalmers University of Technology. Throughout his academic career, he has been endeavoring to practically improve transport safety, efficiency, equity, and sustainability through traffic flow modelling, network optimization, and most recently emerging technologies. In particular, his research has been applied to improvement of emergency services, operations of electric vehicles and connected automated vehicles, and management of vulnerable road users. He has authored or co-authored over 90 journal articles published at top tier journals in the area of transport engineering, and he is a recipient of many prestigious awards. His research has been supported by Australian Research Council Discovery Programme, Queensland Department of Transport and Main Roads, Sydney Trains, National Natural Science Foundation of China, Swedish Innovation Agency Vinnova, and European Union. Xiaobo is now an Associate Editor/Editorial Board Member for IEEE Trans on Cybernetics, IEEE ITS Magazine, Journal of Transportation Engineering ASCE, Transportation Research Part A, Computer-Aided Civil and Infrastructure Engineering, etc.

Dr. Lu Zhen got his Ph.D. from Shanghai Jiaotong University. He has 10 years of teaching experience and 7 years of corporate experience. Dr. Lu Zhen got his Ph.D. in 2008 from Shanghai JiaoTong University. During 2008–2010 he made the Post-Doctoral Research Fellow in National University of Singapore. His research interests lie mainly in Decision Support Systems; Knowledge Management; Information Systems; Optimization and Simulation; Port Operations.

Dr. Robert J. Howlett is the Executive Chair of KES International, a non-profit organization that facilitates knowledge transfer and the dissemination of research results in areas including Intelligent Systems, Sustainability, and Knowledge Transfer. He is a Visiting Professor at Bournemouth University in the UK. His technical expertise is in the use of intelligent systems to solve industrial problems. He has been successful in applying artificial intelligence, machine learning and related technologies to sustainability and renewable energy systems; condition monitoring, diagnostic tools and systems; and automotive electronics and engine management systems. His current research work is focussed on the use of smart microgrids to achieve reduced energy costs and lower carbon emissions in areas such as housing and protected horticulture.

Lakhmi C. Jain, B.E.(Hons), M.E., Ph.D. Fellow (IE Australia) is with the Faculty of Education, Science, Technology & Mathematics at the University of Canberra, Australia and the University of Technology Sydney, Australia. He is a Fellow of the Institution of Engineers Australia. Professor Jain founded the KES International for providing a professional community the opportunities for publications, knowledge exchange, cooperation and teaming. Involving around 5000 researchers drawn from universities and companies worldwide, KES facilitates international cooperation and generates synergy in teaching and research. KES regularly provides networking opportunities for professional community through one of the largest conferences of its kind in the area of KES (www.kesinternational. org). His interests focus on the artificial intelligence paradigms and their applications in complex systems, security, e-education, e-healthcare, unmanned air vehicles and intelligent agents.

Chapter 1
An Algorithm for Reducing Vehicles' Stop Behind the Bus Pre-signals

Mina Ghanbarikarekani, Michelle Zeibots and Yun Zou

Abstract One of the current controversial issues is mitigating traffic congestion in big cities. Public vehicles play a vital role in solving this matter, so it has been taken into consideration to improve not only the public transit's infrastructure but also its functionality. More specifically, it is aimed to encourage travellers to use public vehicles. In order to serve this purpose, public transit systems' delay needs to be decreased by prioritising them. Nowadays, pre-signal, as a modern strategy for prioritising buses, has been proposed. Pre-signal is an additional signal that is applied near the intersection to give buses priority through stopping vehicles in advance of the main intersection, and it reduces the average delay per traveller. However, it has to be mentioned that pre-signals penalise private vehicles by extra stops. This paper aims to propose a model to improve the pre-signal strategy by reducing the vehicles' number of stops behind the pre-signals. This model would cause vehicles to be able to adjust their speed based on traffic conditions and traffic signal as well as buses' speed and approach.

1.1 Introduction

Traffic congestion is one of the most significant issues in big cities, so suggesting methodologies, which eliminate traffic congestion, would pave the way for improving transportation systems in main cities, specifically in their central business district [1–3]. Improving the functionality of public transit systems would reduce traffic jam, and prioritising these systems is one of the influential methods in this target [4–7]. During the past two decades, implementing pre-signals in advance of the signalised

M. Ghanbarikarekani (✉) · M. Zeibots · Y. Zou
School of Civil and Environmental Engineering, University of Technology Sydney, Sydney, NSW 2007, Australia
e-mail: Mina.ghanbarikarekani@student.uts.edu.au

M. Zeibots
e-mail: Michelle.e.zeibots@uts.edu.au

Y. Zou
e-mail: Yun.zou@student.uts.edu.au

© Springer Nature Singapore Pte Ltd. 2019
X. Qu et al. (eds.), *Smart Transportation Systems 2019*, Smart Innovation,
Systems and Technologies 149, https://doi.org/10.1007/978-981-13-8683-1_1

1

intersections has been proposed in order to prioritise buses [8]. Applying pre-signals upstream of the signalised intersections results in decreasing the discharge rate of the intersections [9–13]. Several solutions proposed for solving the problem of low discharge rate at intersections, such as queue relocation [10], mid-block pre-signals [13], an adaptive algorithm that controls pre-signals regarding real-time demand for private and public transportation [11], tandem design [12].

Extending the queue length to the upstream intersection is another pre-signals' impact [9]. Moreover, Kejun [14] carried out a research, and demonstrated pre-signals worsen the private vehicles' performance due to additional stops of private vehicles behind pre-signals. For solving this issue, He et al. proposed a real-time controlling algorithm [15]. Guler et al. performed a dynamic procedure of timing the pre-signals for decreasing private vehicles' delay in undersaturated intersections and increasing their discharging rate in oversaturated intersections [16].

Based on previous studies conducted on pre-signals, these systems could cause additional stops, delay and travel time for private vehicles. This paper aims to solve the mentioned problem as well as provide priority for buses by minimising the cars' time-in-queue behind the pre-signal through modifying their speed while reaching the intersection.

1.2   Research Methodology

In this paper, we propose a model for improving the performance of the intersections equipped with pre-signals. The model is based on minimising the time-in-queue per vehicle behind the pre-signal. To this end, it needs to estimate an appropriate speed for arriving vehicles with the minimum time that they spend in the queue behind pre-signals. More specifically, it would be necessary to minimise the vehicles' time-in-queue based on the equation presented in the Highway Capacity Manual (HCM) [17].

Time-in-queue has been suggested to measure the delay of vehicles directly instead of calculating control delay per vehicle and progression adjustment factor. In order to calculate this parameter, it needs to estimate vehicle-in-queue, and substitute it in Eq. (1.2). It is equal to counts of vehicles approaching the intersection when the signal is green, and the front vehicles have stopped [17].

It requires to present some assumptions for explaining the algorithm: the considered intersection is isolated, and its pre-signalised approach is a one-way two-lane road, the intersection is controlled by a fixed-time traffic signal, a Variable Message Sign (VMS) is installed for alarming private vehicles, the Automatic Vehicle Location system (AVL) is embedded in each bus for the sake of determining the buses' location, their approach and selected lane, detectors are needed throughout the area behind the stop line where buses can change their lane, a detector is needed to count the cars occupied the distance between the VMS and the area behind the stop line. Objective function

$$F(x) = \min [d_{vq}]$$ (1.1)

Subject to

$$d_{vq} = \left(I_s \times \frac{\sum\limits_{i=1}^{n} \sum\limits_{j=1}^{m} V_{ijq}}{V_{tot}} \right) \times 0.9$$ (1.2)

$$v_{car,suggested} \geq 5 \text{ km/h}$$ (1.3)

In Eq. (1.2), d_{vq} is the time-in-queue per vehicle (s), I_s is the interval between vehicle-in-queue counts (veh), $\sum_{i=1}^{n} \sum_{j=1}^{m} V_{ijq}$ is the sum of vehicle-in-queue counts (i) in all lanes (j) (veh), V_{tot} is the total number of vehicles arriving during the survey period (veh), and 0.9 is the empirical adjustment factor.

The next phase is estimating the appropriate speed to be suggested to the private vehicles arriving at pre-signals. To this end, the optimal distance for applying pre-signals (d_{bus}), and the distance of installing a Variable Message Sign (VMS) for informing the cars about their speed (d_{VMS}) have to be calculated using the Eqs. (1.4) and (1.5) proposed by Ghanbarikarekani et al. [1]. Their research aims to propose a model for improving the pre-signal operation through minimising initial queue delay behind pre-signals at one-lane roads.

$$d_{bus} = \left[(c - r_{ms})_1 \times \left(\frac{C}{k_{jam}} - v_{bus,1} \right) + (c - r_{ms})_2 \times \left(0 - \frac{C}{k_{jam}} \right) \right] \times 1000$$ (1.4)

In the above-mentioned equation, d_{bus} is the distance between the pre-signal and the main intersection (m), c is the common cycle length of the traffic signal (h), r_{ms} is the red duration at the traffic signal (h), C is the total capacity across all lanes at the main signal (veh/h), k_{jam} is the jam density (density at zero speed) (veh/km), $\frac{C}{k_{jam}}$ is the bus moving speed for changing lane (for doing right turn) and reaching the intersection, $v_{bus,1}$ is the bus initial speed (which is not zero), $(c - r_{ms})$ is the green time, $(c - r_{ms})_1$ is the required time for buses to change the lane with initial speed of v_1, and $(c - r_{ms})_2$ is the required time for buses to reach the stop line with speed of $\frac{C}{k_{jam}}$.

$$x = \frac{v_{current} t_{reaction}}{3.6}$$ (1.5)

$$d_{VMS} = x + d_{bus}$$ (1.6)

In this equation, $v_{current}$ is the speed of the lead vehicle, which is assumed constant and based on the road rules (km/h), $t_{reaction}$ is the reaction time of the lead vehicle once decides to change its speed (s), which is assumed 2 s, x is the distance between

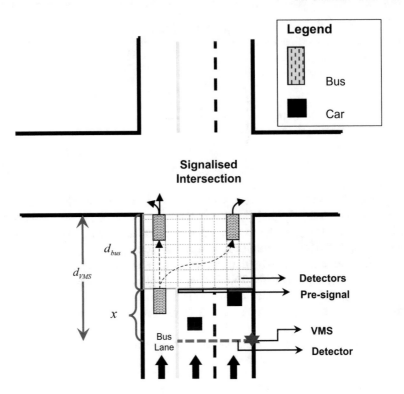

Fig. 1.1 The schematic intersection components

the lead vehicle and the pre-signal, d_{VMS} is the essential distance between the lead vehicle and the intersection which is required for the lead vehicle in the critical situation to stop the car (m) (Shown in Fig. 1.1).

Using d_{bus} and d_{VMS}, the appropriate speed for private vehicles could be calculated. This speed would be optimised by minimising the amount of time-in-queue for each vehicle. In other words, the time-in-queue of the vehicles arriving the pre-signal is to be reduced through adjusting their speed based on the arrival time of the bus and traffic condition prior to the intersection. Moreover, they need to be informed about their operating speed by VMS. Hence, they would be able to move with the suggested speed from VMS point to pre-signal for the sake of minimising their time-in-queue behind the pre-signal.

In order to estimate the speed of private cars, the equations proposed by Ghanbarikarekani et al. [1] are applied.

$$v_{car,suggested} = \frac{d_{VMS}}{t_C} \times 3.6 \tag{1.7}$$

While $v_{car,suggested}$ is the suggested speed of the car in arterials according to the leading traffic condition, in terms of the existence of the buses changing their lanes and traffic signal timing (km/h). Moreover, t_C is the car's travel time according to the leading traffic condition (s) which could be calculated by Eq. (1.8).

$$t_C = t_B + t_{car} \tag{1.8}$$

t_B is the bus's travel time according to traffic signal timing (s), and t_{car} is the car's travel time to reach the intersection (s), which are computed through Eqs. (1.9), (1.10), respectively.

$$t_B = Max\ [t_{bus}, r_{ms}] \tag{1.9}$$

$$t_{bus} = \frac{3.6 d_{bus}}{v_{bus}} \tag{1.10}$$

$$t_{car} = \frac{3.6 d_{VMS}}{v_{car,current}} \tag{1.11}$$

t_{bus} is the bus's travel time (for changing its lane) to pass the area behind the stop line (s), d_{bus} is the distance between the pre-signal and the main intersection (m), v_{bus} is the bus's mean speed of near the intersection (km/h), r_{ms} is the duration of red time-in-traffic signal (s). $v_{car,current}$ is the car's mean speed in arterials (km/h), and d_{VMS} is the proposed distance for installing the VMS (m).

1.3 Results and Discussion

The sensitive analysis of the proposed model would be done in this section in order to investigate the efficiency of applying the algorithm. To serve this purpose, we use the hypothetical value presented in Table 1.1. Most of these parameters such as cycle length and capacity have been assumed.

Using this data the required parameters could be determined, such as d_{bus}, d_{VMS}, $v_{car,suggested}$ and d_{vq}. Then, time-in-queue would be estimated before and after implementing the proposed model. In the end, the efficiency of the algorithm would be investigated by comparing the time-in-queue value in the current situation with its value after utilising the suggested algorithm. This comparison is shown in Fig. 1.2.

As it is shown in Fig. 1.2, implementing the proposed model has declined time-in-queue at all V/C ratios. It has to be mentioned that time-in-queue is decreasing remarkably in higher V/C ratios. In other words, in oversaturated traffic condition, time-in-queue has been reduced dramatically in comparison with undersaturated conditions.

Table 1.1 Hypothetical values of parameters used in the model

Parameters	Values (current situation)	Values (after implementing the model)
Number of cars (N)	800 veh	800 veh
Number of buses	20 veh	20 veh
Car headway	4.5 s	4.5 s
Cycle length (c)	60 s	60 s
Red duration (r_{ms})	30 s	30 s
Capacity (C)	800 veh/h	800 veh/h
Jam density (K_{jam})	100 veh/km	100 veh/km
Initial speed of bus ($V_{bus,1}$)	20 km/h	20 km/h
Initial speed of car ($V_{car,1}$)	40 km/h	40 km/h
Green duration (G)	30 s	30 s
Time for bus to change lane	15 s	15 s
Time for bus to reach the stop line	15 s	15 s
Capacity/density	8	8
d_{bus}	83.33 m	83.33 m
Distance of VMS for car (x)	–	22.22 m
Reaction time (t)	2 s	2 s
d_{VMS}	–	106 m
Bus travel time (t_{bus})	15 s	15 s
Suggested travel time of bus (t_B)	30 s	30 s
Car travel time (t_{car})	–	9.5 s
Suggested travel time of car (t_C)	–	39.5 s
Suggested speed of car ($V_{car,suggested}$)	–	10 km/h
Traffic flow (V)	800 veh/h	800 veh/h
X (V/C)	1.00	1.00
Interval between vehicles in queue I_s	4.5 s	4.5 s
Sum of vehicle-in-queue V_{iq}	640 veh	154 veh
Total number of vehicles arriving during the survey V_{tot}	200 veh	200 veh
Empirical adjustment factor 0.9	0.9	0.9
Time-in-queue d_{vq}	12.96 s	3.12 s
g/c	0.5	0.5

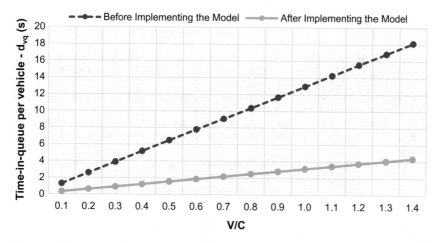

Fig. 1.2 Comparison of each vehicle's time-in-queue before and after implementing the proposed model in different V/C ratios

1.4 Conclusion

Prioritising public vehicles at signalised intersections is one of the effective policies in mitigating traffic congestion through motivating passengers not to use their private cars. Pre-signal is an innovative strategy for prioritising buses by giving the red signal to cars in advance of the main intersection. Hence, cars receive additional stop, and face a higher delay.

This paper suggested a mathematical algorithm in order to improve the cars' performance at signalised intersections equipped with pre-signal. In this paper, it is considered to minimise the private vehicles' time-in-queue behind the pre-signal. To serve this purpose, the appropriate speed of cars, approaching the pre-signal, would be determined to have minimum time-in-queue, delay, and stop time.

According to the results mentioned in the previous section, the proposed model has reduced time-in-queue of the private vehicles behind the pre-signal, so their functionality has been improved remarkably. It needs to be considered that the effect of this model on higher V/C ratios is more than smaller V/C ratios. In other words, this algorithm could improve traffic performance of cars in oversaturated traffic conditions more efficiently than undersaturated ones.

References

1. Ghanbarikarekani, M., Qu, X., Zeibots, M., Qi, W.: Minimizing the average delay at intersections via pre-signals and speed control. J. Adv. Transp. **2018**, 1–8 (2018). https://doi.org/10.1155/2018/4121582. Article ID 4121582, 8 pp.
2. Li, X., Medal, H., Qu, X.: Connected heterogeneous infrastructure location design under additive service utilities. J. Transp. Res. Part B **120**, 99–124 (2019)
3. Qu, X., Zhang, J., Wang, S.: On the stochastic fundamental diagram for freeway traffic: model development, analytical properties, validation, and extensive applications. J. Transp. Res. Part B **104**, 256–271 (2017)
4. Wang, S., Qu, X.: Station choice for Australian commuter rail lines: equilibrium and optimal fare design. Eur. J. Oper. Res. **258**(1), 144–154 (2017)
5. Wang, S., Zhang, W., Qu, X.: Trial-and-error train fare design scheme for addressing boarding/alighting congestion at CBD stations. Transp. Res. Part B **118**, 318–335 (2018)
6. Liu, Z., Wang, S., Chen, W., Zheng, Y.: Willingness to board: a novel concept for modeling queuing up passengers. J. Transp. Res. Part B **90**, 70–82 (2016)
7. Liu, Z., Chen, X., Meng, Q., Kim, I.: Remote park-and-ride network equilibrium model and its applications. J. Transp. Res. Part B **117**, 37–62 (2018)
8. Peake, M.: Getting buses to the front of the queue. In: IPENZ Transportation Conference 2006, Queenstown, New Zealand (2006)
9. Wu, J., Hounsell, N.: Bus priority using pre-signals. Transp. Res. Part A: Policy Pract. **32**(8), 563–583 (1998)
10. Kumara, S.S.P., Hounsell, N.B.: Bus priority and queue relocation capabilities of pre-signals: an analytical evaluation. In: 11th IFAC Symposium on Control in Transportation Systems, Delft, The Netherlands, pp. 391–396 (2006)
11. He, H., Guler, S.I., Menendez, M.: Adaptive control algorithm to provide bus priority with a pre-signal. J. Transp. Res. Part C: Emerg. Technol. **64**, 28–44 (2016)
12. Xuan, Y.: Increasing the flow capacity of signalized intersections with pre-signals: theory and case study. Ph.D. thesis, University of California, Berkeley (2011)
13. Xuan, Y., Cassidy, M.J., Daganzo, C.F.: Using a pre-signal to increase bus- and car- carrying capacity at intersections: theory and experiment. In: TRB 2012 Annual Meeting (2012)
14. Kejun, L.: Bus priority signal control at isolated intersection. In: International Conference on Intelligent Computation Technology and Automation 2008, IEEE Computer Society, Montpellier, France, vol. 2, pp. 234–237 (2008)
15. He, H., Guler, S.I., Menendez, M.: Providing bus priority using adaptive pre-signals, J. Online pre-signal strategies TRB - final after reviews (2014). http://docs.trb.org/prp/15-2793.pdf
16. Guler, S.I., Gayash, V.V., Menendez, M.: Bus priority at signalized intersections with single-lane approaches: A novel pre-signal strategy. J. Transp. Res. Part C: Emerg. Technol. **63**, 51–70 (2016)
17. Highway Capacity Manual: Metric Units 2000, Transportation Research Board. DC, the United States of America, Washington (2000)

Chapter 2
Estimation Models for the Safety Level of Indoor Space Pedestrian Flows

Zewen Wang, Renwei Liu, Xinhua Wu and Zhiyuan Liu

Abstract This paper presents a method to safely organize the indoor space pedestrian flows. To quantitatively describe the indexes related to pedestrian safety, a capacity estimation model is proposed for the number of participants, and then the characteristic of pedestrians of different age is considered in analysing pedestrian speed by using step frequency data. Furthermore, we develop a pedestrian safety estimation model and obtain the probability of safe crossing. Based on the estimation results of capacity model and safety estimation model, a reasonable scheme of pedestrian organization is put forward. Finally, a case study is adopted to apply the proposed models.

2.1 Introduction

To guarantee the safe and efficient operation of urban multimodal transport systems is of essential significance for the mega metropolitans [1, 2]. Previous studies on transport safety mainly focus on the urban road network, freight logistics and public transport systems. Emerging needs are observed for the safety studies in the sector of pedestrian [3, 4]. It is necessary to quantitatively assess the operation of pedestrian flows. Most of the existing pedestrian-related studies focus on the terminals or stations, such as bus stations, subway stations, railway stations, etc. [5–8]. Fruin [9] investigated the pedestrian walking speed, traffic characteristics, pedestrian queuing characteristics, space and conflict; In the HCM, the speed–density relationship of pedestrian traffic flow under different travel purposes is studied [10]; Sisiopiku and Akin [11] investigated pedestrian traffic behaviour and psychological feelings at various traffic facilities. Helbing [12] proposed a dynamic-based social force model, which describes the sociality of pedestrian walking from the force perspective. Blue and Adler [13] established the classical cellular automata model and

Z. Wang · R. Liu · X. Wu · Z. Liu (✉)
Jiangsu Key Laboratory of Urban ITS, Jiangsua Province Collaborative Innovation Center of Modern Urban Traffic Technologies, School of Transportation, Southeast University, Nanjing, China
e-mail: zhiyuanl@seu.edu.cn

© Springer Nature Singapore Pte Ltd. 2019
X. Qu et al. (eds.), *Smart Transportation Systems 2019*, Smart Innovation, Systems and Technologies 149, https://doi.org/10.1007/978-981-13-8683-1_2

applied it to pedestrian traffic flow research. After that, Ren et al. [14] established the CA pedestrian simulation model and analysed the self-organization phenomenon in the simulation process.

With the popularization of mobile GPS and the coming of the era of wide application of big data, Nikolicm et al. [15] put forward a probability model of pedestrian traffic flow velocity–density, describing the relationship between pedestrian speed density from the perspective of probability. Hoogendoorn and Bovy [16] studied the characteristics and basic parameters of pedestrian flow at one-way, opposite and bottleneck through experiments, and established a pedestrian behaviour model, NOMAD model. Yang [17] and Xing [18] studied the pedestrian traffic characteristics inside the passenger terminal and determined the basic parameters of various facilities of large passenger terminal. Lam et al. [19] analysed the data of pedestrian speed and flow in Hong Kong commercial blocks and drew its basic relationship map. Zhu et al. [20] studied the service capability of pedestrian facilities to meet the needs of large-scale participants at different stages of pedestrian traffic flow operation.

The pedestrian demand of terminals or stations is rigid and may not be limited, while indoor space, such as science and technology museum, has flexible demand, which can restrict pedestrian flow from the source. Additionally, a large number of pedestrians tend to arrive and leave the stadium at the same time when the events start and end, forming a massive accumulation in space in a short period. Pedestrians of the indoor space studied in this paper have the characteristics of random arrival and departure because of its exhibition function. Therefore, it is not always applicable to use traditional models to describe the state of indoor pedestrians and more suitable method is deemed necessary.

The structure of this paper is as follows. Section 2 explains the method used in this paper to estimate the capacity of indoor space and safety degree of a pedestrian when leaving the indoor space. Section 3 presents a case of study using the proposed models. Section 4 provides a conclusion of this paper and recommendations.

2.2   Estimation Models

This chapter introduces the main method used in this study. First, the reasonable per capita occupancy area is determined according to the different behaviour patterns of pedestrians in various spaces of the indoor space, and the reasonable audience capacity of the indoor space is obtained. Then, the pedestrian traffic safety is evaluated by using the pedestrian crossing safety degree model.

2.2.1   Estimation of Capacity of Indoor Space

According to Tan [21], the space in the exhibition hall can be divided into three parts: ornamental space, rest space and traffic space. The reasonable occupied area for per viewer is different due to the different functions of each space. Based on this principle, the key to the calculation of the reasonable use of the entire indoor space is to determine the per capita area index. Therefore, if per capita area of participants in each type of space is determined, the reasonable number of participants in the museum should be the sum of three parts to reasonably accommodate the number of participants. The basic expressions are as follows:

$$P = P_1 + P_2 + P_3 = \frac{S_1}{S_{i1}} + \frac{S_2}{S_{i2}} + \frac{S_3}{S_{i3}} \tag{2.1}$$

where P is the reasonable number of participants in the exhibition hall; P_1, P_2, P_3 are the number of participants in ornamental space, rest space and traffic space, respectively; S_1, S_2, S_3 are the net area of ornamental space, rest space and traffic space, respectively, and S_{i1}, S_{i2}, S_{i3} are per capita area of participants for ornamental space, rest space and traffic space, respectively.

2.2.2 Estimation Model for Pedestrian Crossing Safety Degree

Pedestrians arriving and leaving the indoor space inevitably have conflicts with vehicles running on the road connected to the indoor space. In this paper, the expected waiting time of pedestrians is defined as the maximum time that pedestrians can accept and willing to wait subjectively and the risk degree of the pedestrian crossing is modelled.

The pedestrian crossing risk model based on safety degree proposed by Xiang and Zhang [22] takes both the arrival of motor vehicles on the road and the demand of pedestrians crossing the street into account, which achieves the quantification of pedestrian crossing safety. It comprehensively reflects the stochastic characteristics of the arrival of motor vehicles or pedestrians and predicts that the actual waiting time of pedestrians crossing the street is less than the probability equal to the expected waiting time, that is, the probability of pedestrian crossing safety. The revised safety model is as follows due to the independence of pedestrians and vehicles arrival:

$$F(z) = P(T' < T) = \frac{\iiint\limits_{H'-P \leq T} f_{H'}(h') f_P(p) f_T(t) dh' dp dw \cdot \iint\limits_{H-P < t_0} f_H(h) f_P(p) dh dp}{\iint\limits_{H-P < t_0} f_H(h) f_P(p) dh dp}$$

$$+ \iint\limits_{H-P \geq t_0} f_H(h) f_P(p) dh dp \tag{2.2}$$

where T' and T are actual waiting time and expected waiting time for pedestrian crossing the road; H' is the time interval between the last acceptable clearance and the next acceptable interval obtained by pedestrian; P is represents the time interval of pedestrian arrival; let H presents the minimum headway time of all lanes in a section and t_0 presents the usual time for pedestrians to cross the road. Also, let $f_{H'}(\cdot)$, $f_H(\cdot)$, $f_P(\cdot)$, $f_W(\cdot)$ denote the probability density function of pedestrian acceptable time interval, minimum headway time, pedestrian arrival time interval and expected waiting time for pedestrian crossing the road, respectively. t_0 can be calculated by the following model:

$$t_0 = \frac{W}{v_0} + t_q \tag{2.3}$$

where W is pavement width; v_0 is the average speed of pedestrian and t_q is reaction time of pedestrians.

2.3 A Case Study of Safety Exhibition Museum

2.3.1 Data Collection

Ruyixi safety experience museum in southern Changchun, China, where connected with the entrance and exit of the indoor space is a two-way two-lane road with a width of 9 m and no central separation facilities. The location map and the indoor space structure map are shown in Figs. 2.1 and 2.2, respectively.

The pedestrian volume and traffic volume were collected during the morning (9:00 am–12:00 pm) and afternoon (3:00 pm–6:00 pm) on the weekend. Direct observation method was adopted and recorded the pedestrian volume and the traffic volume every 15 min, which is contributed to calculating the distribution of the data. Data such as indoor space area come from its planning map. The data associated with pedestrian step length and frequency was also obtained by direct observation, and the size of the sample was 400. The two-way traffic volume of this section is about 850 vehicles per hour, and the pedestrian traffic volume of the exit is about 180 people per hour.

Fig. 2.1 Location map of Ruyixi safety experience museum

Fig. 2.2 Indoor space structure map

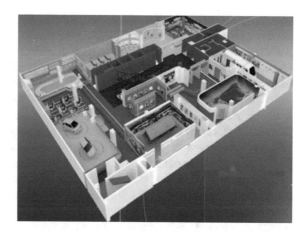

2.3.2 Calculation of Reasonable Participants Number in Indoor Space

(1) Reasonable space demand per capita in ornamental space

Considering the interaction of the facilities of the safety experience museum, the basic behaviour pattern of the participants in the exhibition area is different from those in the museum, so the viewing space S_{i1} is close to 0. Therefore, the reasonable number of viewers in viewing space P_1 cannot be represented by S_1/S_{i1}. Assuming that an exhibit can only be operated by one person, the sum of the number of people in the space, P_1, is approximately the number of exhibits.

(2) Reasonable space demand per capita in rest space

There are no seats in the rest space of the safe experience museum, and the audience is scattered. At this time, the participants are basically in a static state. Pedestrian demand for rest activities is that their basic activities are not disturbed. Referring to the human body size of Chinese adults (GB/T10000-1988), we can know that when the arm stretching length is 1.80 m, it can meet the rest space needs of more than 96% of the adult population. However, due to the large proportion of children in the group, pedestrians are simplified to a cylinder with a radius of 0.8 m, so the reasonable per capita area S_{i2} at this time is 2.00 m^2 approximately.

(3) Reasonable space demand per capita in traffic space

The participants in the safety experience museum are mostly companionate, and the main behaviour pattern of these pedestrians in traffic space is non-directional. From the perspective of individual participant, their movements have specific direction and purpose. Assuming that two people are walking side by side, the common occupancy distance is equivalent to 2.5 shoulder width. According to the proportion of human body, the extension length of two arms is about height, and the shoulder width is about a quarter of height. Therefore, the minimum distance to keep pedestrians from touching each other is 0.45 m. Based on the pedestrian speed survey, the average walking speed of adults is 1.2 m/s. Due to the influence of large proportion of children participants, the indoor space pedestrian speed decreases slightly, set to 1.0–1.1 m/s, reaction time is 0.15–0.4 s. If both of them take a certain value in this range, the pedestrians in the rear will avoid the pedestrian meeting in front by 0.5 m, and then from the perspective of front pedestrians, at least $0.5 + 0.45 = 0.95$ m is needed for non-contact area. Therefore, to guarantee that pedestrians walk at normal speeds without conflicting with each other, a radius of 0.95 m is chosen, in which case the per capita area of the participants S_{i3} is 2.83 m². According to the basic data of the initial stage of the building, the total public building area of the indoor space is 7300 square metres. Combining with the proportion of various functional areas of similar indoor space, it is estimated that the exhibition area covers 3100 m², the traffic area is 3000 m², and the rest area is 1200 m². Assuming that an experience facility or exhibit is displayed every 15 m², the total number of exhibits P_1 is estimated to be 207. Substituting the above results into Eq. (2.1), the total number of participants of indoor space is 1867.

2.3.3 Estimation of Pedestrian Safety Degree

Pedestrian walking speed is related to age, gender and whether there are obstacles in walking area. It was found that pedestrian step length and frequency were also affected by similar factors. Therefore, it is reasonable to think that there is a certain relationship between pedestrian walking speed and step speed characteristics. It is possible to predict the walking speed of pedestrians by obtaining pedestrian step frequency data, which are easier to observe compared with walking speed. According to the regression result of each pair of variables, we select the most relevant variable to predict pedestrian speed, which is shown as the figure.

As shown in Fig. 2.3 and Eq. (2.4), the logarithmic regression model maps pedestrian speed as a function of step frequency, thus there is a corresponding speed for

Fig. 2.3 Regression result of pedestrian speed and step frequency

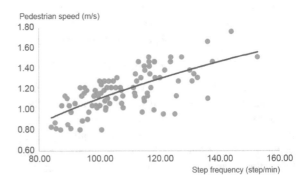

Table 2.1 The weighted average of pedestrian around the indoor space

Age	Proportion (%)	Average speed (m/s)	Weighted average (m/s)
Old age	20	0.94	1.09
Middle age	43	1.18	
Children	37	1.06	

Table 2.2 Pedestrian expected waiting time survey

Waiting duration/s	Middle value	Frequency
0–10	5	0.08
10–20	15	0.26
20–30	25	0.35
30–40	35	0.16
40–50	45	0.09
50–60	55	0.06

each determined step frequency, which is the idea of walking speed estimation. Combined with the gender and age proportion of pedestrians around the indoor space in Table 2.1, the average walking speed can be finally obtained by weighted average, which can be used as the input variable in next section.

$$y = 1.0695 \ln(x) - 3.814 \tag{2.4}$$

The characteristic of pedestrians of different age should be considered in predicting pedestrian speed based on step frequency data, which is an important parameter in pedestrian safety estimation model.

As pedestrian reaction time t_q always equals to 1 s, pedestrian crossing time can be calculated by

$$t_0 = \frac{W}{v_0} + t_q = \frac{9}{1.09} + 1 = 9.26 \, \text{s} \tag{2.5}$$

Using the data in Table 2.2 to fit the lognormal distribution, the following probability density $f_T(t)$ are obtained.

$$f_T(t) = \frac{1}{\sqrt{2\pi} \times 0.56} e^{\left(-\frac{(\ln t - 3.09)^2}{2 \times 0.56^2}\right)} \tag{2.6}$$

Pedestrian arrival obeys Poisson distribution, so pedestrian arrival time interval obeys negative exponential distribution. The probability density function is as follows.

$$f_P(p) = \frac{180}{3600} \times e^{-\frac{180}{3600}p} \tag{2.7}$$

Similarly, the probability density function of vehicles is

$$f_H(h) = \frac{850}{3600} \times e^{-\frac{850}{3600}p} \tag{2.8}$$

Not all gaps between vehicles can be accepted by pedestrians. In fact, only when the headway H of two vehicles is larger than the expected crossing time of pedestrians t_0 the headway is considered acceptable. Considering the interval between two consecutive acceptable time intervals still obeys the negative exponential distribution, the probability density function of acceptable time interval can be obtained by 3 steps, which are described as follows:

Step 1: Calculating the probability $P(H > t_0)$

$$P(H > t_0) = 0.11 \tag{2.9}$$

Step 2: Calculating the new arrival rate λ'

$$\lambda' = \frac{Q}{3600} \cdot P(H > t_0) = 0.026 \tag{2.10}$$

Step 3: According to additive characteristic of Poisson distribution, the probability density function of the time interval between the two consecutive acceptable intervals can be obtained by

$$f_{H'}(h') = \lambda' \cdot e^{-\lambda' \cdot h'} \tag{2.11}$$

From Eq. (2.2), the safety degree of pedestrian crossing the road is 58% approximately.

According to the estimation capacity of the indoor space mentioned above, it is possible to carry out dynamic monitoring of the flow in and out of the indoor space, and the number of indoor space participants can be limited in real time. When the pedestrian number is approaching the capacity, the time limit policy for visiting can be implemented. Also, the low level of pedestrian safety degree means a high risk for pedestrian crossing, and the ideal waiting time of pedestrians has an important impact on the safety of crossing the street. Even under the condition of signal control, most pedestrians crossing the street are still more likely to run through the red light. Therefore, on the way for pedestrians to leave the indoor space, warning signs can be set up and electronic screens can be installed to display accident data for pedestrians and publicize traffic safety.

2.4 Conclusion

This paper introduced a method to safely organize the indoor space pedestrian flows based on the safety level concept. Based on the estimation result, some strategies were put forward for the organization of pedestrian traffic in indoor space. From the analysis of this paper, it is necessary to quantitatively describe the safety level of indoor space pedestrian flow to find the risk factors. After that, a pedestrian organization scheme can be formulated appropriately. This study can provide a good idea for the future pedestrian organization design of similar indoor space.

Acknowledgements This study is supported by the General Projects (No. 71771050) and Key Projects (No. 51638004) of the National Natural Science Foundation of China.

References

1. Liu, Z., Chen, X., Meng, Q., Kim, I.: Remote park-and-ride network equilibrium model and its applications. Transp. Res. Part B **117**, 37–62 (2018)
2. Wang, S., Zhang, W., Qu, X.: Trial-and-error train fare design scheme for addressing boarding/alighting congestion at CBD stations. Transp. Res. Part B Methodol. **118**, 318–335 (2018)
3. Kuang, Y., Qu, X., Wang, S.: A tree-structured crash surrogate measure for freeways. Accid. Anal. Prev. **77**, 137–148 (2018)
4. Liu, Z., Wang, S., Chen, W., Zheng, Y.: Willingness to board: a novel concept for modeling queuing up passengers. Transp. Res. Part B **90**, 70–82 (2016)
5. Gu, Z., Liu, Z., Nirajan, S., Yang, M.: Video-based analysis of school students' emergency evacuation behavior in earthquakes. Int. J. Disaster Risk Reduct. **18**, 1–11 (2016)
6. Wang, C., Xu, C.C., Dai, Y.L.: A crash estimation method based on bivariate extreme value theory and video-based vehicle trajectory data. Accid. Anal. Prev. **123**, 365–373 (2019)
7. Qu, X., Yang, Y., Liu, Z., Jin, S., Weng, J.: Potential Crash Risks of expressway on-ramps and off-ramps: a case study in Beijing China. Safety Sci. **70**, 58–62 (2014)
8. Wang, C., Xu, C.C., Xia, J.X., Qian, Z.D., Lu, L.J.: A combined use of microscopic traffic simulation and extreme value methods for traffic safety evaluation. Transp. Res. Part C **90**, 281–291 (2018)
9. Fruin, J.: Pedestrian Planning and Design. Metropolitan Association of Urban Designers and Environmental Planners (1996)
10. American transportation research council: Hand of Road Capacity. China Communication Press, Beijing (2007)
11. Sisiopiku, V., Akin, D.: Pedestrian behaviors at and perceptions towards various pedestrian facilities: an examination based on observation and survey data. Transp. Res. Part F **6**(4), 249–274 (2003)
12. Helbing, D.: A mathematical model for the behavior of pedestrian. Behav. Sci. **36**, 298–310 (1991)
13. Blue, V.J., Adler, J.L.: Bi-directional emergent fundamental pedestrian flows from cellular automata micro-simulation. In: International Symposium on Transportation & Traffic Theory (1999)
14. Ren, G., Lu, L., Wang, W.: Modeling of two-way pedestrian flow based on cellular automata and complex network theory. Acta physica sinica **61**(14), 255–264 (2012)
15. Nikolicm, M., Bierlaire, M., Farooq, B., et al.: Probabilistic speed-density relationship for pedestrian traffic. Transp. Res. Part B **89**, 58–81 (2016)

16. Hoogendoorn, S., Bovy, P.: Gas-kinetic modeling and simulation of Pedestrian flows. Transp. Res. Board **1710**, 28–36 (2000)
17. Yang, L.: Study on Pedestrian Traffic Characteristics in Integrated Transport Passenger Hub. Jilin University, China (2009)
18. Xing, Y.: Research on Dynamic Simulation Method of Emergency Evacuation Passenger Flow Organization in Large Passenger Station. Beijing Jiaotong University, China (2010)
19. Lam, W., Morrall, J.F., Ho, H.: Pedestrian flow characteristics in Hong Kong. Transp. Res. Rec. **1487**, 56–62 (1995)
20. Zhu, N., Wang, J., Shi, J.: Application of Pedestrian Simulation in Olympic Experience Hall. Transp. Syst. Eng. Inf. **8**(6), 85–90 (2008)
21. Tan, F.: The method of calculating the reasonable audience number of museums. Huazhong Arch. **26**(1), 95–96 (2008)
22. Xiang, H., Zhang, Q.: Pedestrian crossing risk analysis model for non-signal control section. China Safety Sci. J. **26**(4), 126–130 (2016)

Chapter 3
Station-Level Hourly Bike Demand Prediction for Dynamic Repositioning in Bike Sharing Systems

Xinhua Wu, Cheng Lyu, Zewen Wang and Zhiyuan Liu

Abstract Regarding the issue of unbalanced bike distribution in the bike sharing system (BSS), the accurate prediction of bike demand is of importance to dynamic repositioning. This research focuses on the prediction accuracy of the hourly bike number change on station level. Three frequently used machine learning models, including Random Forest (RF), Gradient Boosting Regression Tree (GBRT) and Neural Network (NN) are trained on the same preprocessed dataset. Meanwhile, two training methods, training a check-in prediction model as well as a check-out prediction model, respectively, and training the model with processed bike number change data directly, are both applied to these three models in order to improve the prediction accuracy. The results show that the method of training on the bike number change data is more effective and the GBRT model trained in this way outperforms other models.

3.1 Introduction

Confronted by the deteriorated congestion and pollutions issues, the transport system has become a big hurdle for the vibrant development of urban cities [1, 2]. Thus, the experts have a consensus for the development of future mobility in urban areas, which is to promote the use of public transport and green travel modes [3, 4]. Recently, the bike sharing system (BSS) is an emerging instrument to facilitate the sustainable development of urban green travel modes. However, the fixed docking stations of the bike sharing system inevitably suffer from the problem of unbalanced distribution. As the bikes at a certain station could be deficient or surplus, the demand for renting

X. Wu · C. Lyu · Z. Wang · Z. Liu (✉)
Jiangsu Key Laboratory of Urban ITS, Jiangsu Province Collaborative Innovation Center of Modern Urban Traffic Technologies, School of Transportation, Southeast University, Nanjing, China
e-mail: zhiyuanl@seu.edu.cn

X. Wu · C. Lyu · Z. Wang
Department of Civil Engineering, Institute of Transport Studies, Monash University, Melbourne, VIC, Australia

© Springer Nature Singapore Pte Ltd. 2019
X. Qu et al. (eds.), *Smart Transportation Systems 2019*, Smart Innovation, Systems and Technologies 149, https://doi.org/10.1007/978-981-13-8683-1_3

or returning at these stations could not be satisfied. The status of stations could be easily monitored while it takes time to transit bikes from one station to another. When a shortage of supply has occurred, there is usually not enough time for reposition. Thus, the prediction of shared bike demand is of importance to benefit the effective repositioning in BSS.

With open operational data of BSS as well as the emerging machine learning technology available, a variety of prediction models have been built [5]. These models could be divided into three spatial levels, namely city-level, cluster-level and station level. The city-level prediction aims to predict the coarse-grained bike usage for all stations in a whole city. Although such a model can forecast the overall future condition of BSS, the over-simplified results are not adequate to guide the repositioning of the system. As to the cluster-level prediction, it assumes that geographically contiguous stations are likely to share similar temporal demand features, thus the total demand of these stations can be predicted as a cluster [6, 7]. Compared with city-level prediction, the cluster-level prediction is able to capture the local demand dynamics, but the peculiarity of stations is still neglected. Therefore, accurate station-level bike demand prediction is necessary for the efficient and prompt repositioning of BSS [8–10]. For this purpose, many previous studies attempted to build prediction models for each station separately. However, such practice fails to make full use of the whole data set and fails to consider the latent relationship and similarity among stations. The performance of the prediction model could be unsatisfying, especially for those newly established stations with limited historical data.

Our research is aimed to build a station-level hourly prediction model which is able to capture the global spatio-temporal features. It should be noticed that the repositioning of BSSs is based on the short-term change of the bike number at each station, more specifically, whether one station will reach its maximum capacity or have no bikes for rent in the subsequent several hours. If the number of bikes check-in or check-out is roughly equivalent to each other at one station, the status of the station can be regarded as balanced. Nevertheless, the dynamic characteristics of BSS are neglected in previous station-level studies, because most of these studies discuss the accuracy of prediction results of check-in and check-out behaviors independently. For example, given two predictive models, Model A and Model B, where the prediction error of Model A is 1 for checking-in and −1 for checking-out at a certain station, while the error of Model B is 2 for checking-in and 2 for checking-out at the same station. In terms of the absolute prediction error of checking-in and checking-out, Model A apparently outperforms Model B, whose error is two times of Model A. However, the result of Model A indicates a 2-bike increase in bike number of this station whereas that of Model B indicates 0, which is closer to the truth that there is no requirement of repositioning at this station. It is the accurate station-level bike number prediction in the near future which is vital for efficient bike repositioning, hence prediction models which tend to overpredict or underpredict the number of check-in and check-out bikes at the same time could also be practically valuable. This paper mainly discusses the prediction accuracy of the hourly bike number change on station level. Three machine learning algorithms, including Random Forest (RF), Gradient

Boosting Regression Tree (GBRT) and Neural Network (NN), are implemented, according to their good performance in previous relevant research.

3.2 Three Models for Bike Demand Prediction

In order to predict the station-level bike number change, there are intuitively two different methods. One is to separately train two sperate prediction models for check-in and check-out. Then, the predicted change of bike number can be easily obtained by calculating the subtraction between the prediction results of these two models. The other method is to calculate the change of bike number first and then train the model with preprocessed bike number change data directly. The bike number change can be calculated as follows.

$$N_{i,t}^{change} = N_{i,t}^{checkin} - N_{i,t}^{checkout} \tag{3.1}$$

where $N_{i,t}^{change}$ represents the change of bike number at station i during the hour t, $N_{i,t}^{checkin}$ represents the check-in bike number at station i during the hour t and $N_{i,t}^{checkout}$ represents the check-out bike number at station i during the hour t.

Three machine learning models including RF, GBRT and NN are adopted and each of them will be trained with the two aforementioned methods. These three machine learning algorithms are briefly introduced as follows.

3.2.1 Random Forest

Random forest is typical decision-tree-based learning algorithm. For a traditional decision tree model, it classifies the instances based on their attributes. Random forest was proposed as an ensemble learning method which combines the results of multiple decision trees [11]. The trees inside the forest are, respectively, constructed using samples bootstrapped from the entire data set with replacement. Moreover, instead of splitting at the best split among all features, the splitting point is determined based on a random subset of features. Finally, for a regression problem such as the bike number prediction, the outputs are obtained by calculating the average of results generated by multiple trees. One advantage of the random forest model is that it does not easily suffer from the over-fitting issue. Besides, it is not computationally costly and easy to implement.

3.2.2 Gradient Boosting Regression Tree

Aside from bagging mentioned in Subsection Random Forest, another frequently used ensemble method is boosting. Rather than directly learning the same objective, boosting learns through a multistep procedure to improve the accuracy of weak learners, where the learners are built on the basis of the previous round of training. Gradient Boosting Regression Tree (GBRT) adopts an additive model to combine multiple base trees based on gradient boosting algorithm [12, 13], which can be formulated as follows:

$$f_K(\mathbf{x}) = \sum_{k=1}^{K} T(\mathbf{x}; \boldsymbol{\theta}_k) \tag{3.2}$$

where $T(x; \theta_k)$ denotes a decision tree with parameters θ_k, K denotes the number of base trees, and k is the index of trees. During training, the negative gradient of the loss function $\mathcal{L}(y_i, T(x_i))$ (which is squared loss in this paper) is taken as an estimate of the residual r_i (see Eq. 3.3), and a new tree can then be built with the updated residual.

$$r_i = -\frac{\partial \mathcal{L}}{\partial T(x_i)} = y_i - T(x_i) \tag{3.3}$$

3.2.3 Neural Network

Artificial Neural Network (NN) is renowned for its ability in tackling non-linear relationships. It has been proved that NN is a class of universal approximators, which are able to learn any mappings from inputs with finite dimensional space to another [14]. A typical NN consists of a vast number of neurons, which attempt to simulate the structure of human brain, and these neurons are connected layer-wise to form a network. Each neuron in the network is affiliated with a weight. By multiplying the vector of a layer of neurons by the corresponding weight matrix, the vector of the next layer can be obtained through the mapping an activation function $f(\cdot)$.

$$X_i = f(W_{i-1} * X_{i-1}) \tag{3.4}$$

Using the backpropagation algorithm, the weights can be updated in each epoch of training.

3.3 Data

3.3.1 Data Description

In this study, the data from Citi Bike, the largest BSS in North America, is used for evaluation of the prediction models [15]. The historical operating data of Citi Bike is open to the public and has benefitted a number of studies relating to bike sharing service since 2013. We extracted the data in June and July 2017 as the training set, and the data in August 2017 as the test set. Considering the limited BSS data in most cities, only 1-month historical data is used in order to improve the generalization capability of the prediction model.

3.3.2 Feature Engineering

The input feature can be categorized into four types, i.e. usage features, spatial features, temporal features and augmented features.

Usage features. Historical bike usage features used in this paper are station specific. Denote the number of pick-ups and drop-offs as $x_t = \{x_{pick,t}, x_{drop,t}\}$, where t is index of hour. To reflect the periodical changes in bike usage, we extracted the bike usage number in the same hour of past 4 weeks $(x_{t-168}, x_{t-336}, x_{t-504}, x_{t-672})$, usage in the past 6 h $(x_{t-1}, x_{t-2}, x_{t-3}, x_{t-4}, x_{t-5}, x_{t-6})$, the total usage number in the past 30 days $\left(\sum_{i=1}^{720} x_{t-i}\right)$, etc.

Spatial features. Spatial features are extracted under two assumptions:

- Citizens choose to rent a shared bike only when the bike station is not far away from them. Since the trip distance of shared bike is relatively short, citizens are not likely to walk to a distant bike station. We choose 500 m as the distance threshold and assume the station will only radiate the area within the range. Figure 3.1 shows the influencing area of the whole BSS.
- Citizens always choose the station which is closest to them. Based on this assumption, we divide the influencing area into Voronoi polygons, which is demonstrated in Fig. 3.2. Every point in a particular Voronoi polygon is closer to the corresponding station than any other stations. As the influencing area of each station is represented by a Voronoi polygon, numerous important spatial features, including area, length of the bicycle lanes and points of interest, can be easily counted for each station.

Temporal features. Temporal features include the hour of a week and the week of a year. Regarding the hour of the week, a week is divided into 168 time slots, each of which represents one hour. The reason for using this feature instead of the hour of a day is that day-to-day usage dynamics are unstable due to the existence of weekends, while the same hour in different weeks shares similar usage patterns.

Fig. 3.1 Influencing area of
the whole BSS

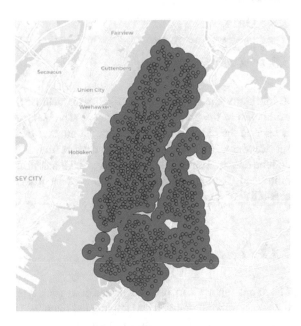

Fig. 3.2 Voronoi polygons
of influencing area

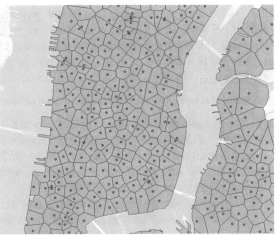

Augmented features. According to previous studies, the feature of weather is vital
to accurate prediction. However, as to a station-level prediction model which aims
to guide the dynamic repositioning, the real-time fine-grained weather information
is hard to obtain hourly. Moreover, when the weather is unfavorable, the bike usage
usually decreases sharply, which indicates that the demand for repositioning of BSS
will be low. In this situation, greater prediction errors could be accepted. Therefore,
only the relatively stable daily mean temperature is added as the augmented feature
in this research.

3.4 Results and Analysis

As mentioned in Sect. 2, there are two methods to predict the station-level bike number change. One of them is to separately train two separate prediction models for check-out and check-in. Following this idea, three machine learning models, i.e. RF, NN and GBRT, are trained on the same dataset, respectively. By calculating the subtraction between the prediction results of these two models, the predicted change number can be easily obtained. The model performance on the test set is listed in Table 3.1, where the evaluation metric is mean absolute error (MAE). Regarding the respective prediction results of check-in and check-out prediction, the GBRT model outperforms the RF and NN model. However, with regard to bike number change prediction, the GBRT model has the highest MAE, 1.9347, indicating that its performance is the worst among the three models. Such results imply that separately trained models may have poor performance when predicting bike number changes, even though the prediction of check-in and check-out is relatively accurate. Meanwhile, it also demonstrates the potential limitations of previous evaluation of prediction models in BSS.

The other method is to train the model on the preprocessed bike number change data directly, and the same three models are applied to this task. Considering the practice of repositioning in BSS, we assume the prediction with an error within 2 bikes as an accurate prediction, which can guide the bike repositioning. Hence, in addition to MAE, a new metric, accurate percentage (AP), which represents the proportion of samples with absolute prediction error under 2 bikes to all samples, is defined to evaluate the model performance. The model with higher AP usually makes more accurate predictions. The performance of these all six models, either separately trained or directly trained, is shown in Table 3.2. The directly trained GBRT model has the best performance among the six models. It can also be observed that the directly trained models all outperform the corresponding separately trained models, which indicates that the method of direct training tends to be more efficient.

Table 3.1 Check-out, Check-in and Change Mae with three machine learning algorithms	Model	Check-out MAE	Check-in MAE	Change MAE
	RF	1.7932	1.7469	1.8682
	NN	1.7990	1.7424	1.8857
	GBRT	1.7280	1.7259	1.9347

Table 3.2 Performance of the six models

Model	Change MAE	AP
RF (separately trained)	1.8682	0.6990
RF (directly trained)	1.8284	0.7052
NN (separately trained)	1.8857	0.6189
NN (directly trained)	1.8830	0.6934
GBRT (separately trained)	1.9347	0.6915
GBRT (directly trained)	1.8159	0.7085

3.5 Conclusions

A number of studies have been focused on the demand prediction in BSS in order to guide the bike repositioning and most of them discuss the prediction accuracy of check-in and check-out behaviors independently. However, as the dynamic repositioning of BSSs is based on the short-term change of the bike number on station level, the prediction accuracy of the bike number change could be more important.

In this paper, three machine learning models are used in order to improve the prediction of the accuracy of the bike number change on station level. Two different methods, separately training two prediction models for check-out as well as check-in, and training the model on the preprocessed bike number change data directly, are both applied to these three models.

The results indicate that directly trained models tend to have better performance, while separately trained models may have poor performance even though the prediction of check-in and check-out is relatively accurate.

Acknowledgements This study is supported by the General Projects (No. 71771050) and Key Projects (No. 51638004) of the National Natural Science Foundation of China.

Commentary Describing the Changes Thanks for the reviewers' helpful advice. This paper has been carefully checked and the format of references has been consistent now. As for the problem of language expression, some vocabulary choices and sentence structures have also been modified in order to avoid potential misunderstanding.

References

1. Liu, Z., Chen, X., Meng, Q., Kim, I.: Remote park-and-ride network equilibrium model and its applications. Transp. Res. Part B Methodol. **117**, 37–62 (2018)
2. Liu, Z., Wang, S., Chen, W., Zheng, Y.: Willingness to board: a novel concept for modeling queuing up passengers. Transp. Res. Part B Methodol. **90**, 70–82 (2016)
3. Liu, Z., Yan, Y., Qu, X., Zhang, Y.: Bus stop-skipping scheme with random travel time. Transp. Res. Part C Emerg. Technol. **35**, 46–56 (2013)
4. Meng, Q., Qu, X.: Bus dwell time estimation at bus bays: a probabilistic approach. Transp. Res. Part C Emerg. Technol. **36**, 61–71 (2013)

5. Li, Y., Zheng, Y., Zhang, H., Chen, L.: Traffic prediction in a bike-sharing system. In: Proceedings of the 23rd SIGSPATIAL International Conference on Advances in Geographic Information Systems—GIS '15, pp. 1–10. ACM Press, Bellevue, Washington (2015)
6. Chen, L., Jakubowicz, J., Zhang, D., Wang, L., Yang, D., Ma, X., Li, S., Wu, Z., Pan, G., Nguyen, T.-M.-T.: Dynamic cluster-based over-demand prediction in bike sharing systems. In: Proceedings of the 2016 ACM. International Joint Conference on Pervasive and Ubiquitous Computing - UbiComp '16, pp. 841–852. ACM Press, Heidelberg, Germany (2016)
7. Feng, S., Chen, H., Du, C., Li, J., Jing, N.: A Hierarchical demand prediction method with station clustering for bike sharing system. In: 2018 IEEE Third International Conference on Data Science in Cyberspace (DSC), pp. 829–836. IEEE, Guangzhou (2018)
8. Lin, L., He, Z., Peeta, S.: Predicting station-level hourly demand in a large-scale bike-sharing network: a graph convolutional neural network approach. Transp. Res. Part C Emerg. Technol. **97**, 258–276 (2018)
9. Chen, P.-C., Hsieh, H.-Y., Sigalingging, X.K., Chen, Y.-R., Leu, J.-S.: Prediction of station level demand in a bike sharing system using recurrent neural networks. In: 2017 IEEE 85th Vehicular Technology Conference (VTC Spring), pp. 1–5. IEEE, Sydney, NSW (2017)
10. Wang, B., Kim, I.: Short-term prediction for bike-sharing service using machine learning. Transp. Res. Procedia. **34**, 171–178 (2018)
11. Breiman, L.: Bagging predictors. Mach. Learn. **24**, 123–140 (1996)
12. Liu, Y., Jia, R., Xie, X., Liu, Z.: A two-stage destination prediction framework of shared bicycles based on geographical position recommendation. IEEE Intell. Transp. Syst. Mag. **11**, 42–47 (2019)
13. Drucker, H.: Improving regressors using boosting techniques. In: Proceedings of the Fourteenth International Conference on Machine Learning, pp. 107–115. Morgan Kaufmann Publishers Inc., Nashville, Tennessee, USA (1997)
14. Friedman, J.H.: Greedy function approximation: a gradient boosting machine. Ann. Stat. **29**, 1189–1232 (2001)
15. Motivate International, Inc.: Citi Bike System Data. https://www.citibikenyc.com/system-data

Chapter 4
Study of Data-Driven Methods for Vessel Anomaly Detection Based on AIS Data

Ran Yan and Shuaian Wang

Abstract Maritime safety and security are gaining increasing concern in recent years. There are a growing number of studies aiming at improving situational awareness in the maritime domain by identifying vessel anomaly behaviors based on the data provided by the Automatic Identification System (AIS). Two types of data-driven methods are most popular in vessel anomaly detection based on AIS data: the statistical methods and machine learning methods. To improve the detection model efficiency and accuracy, hybrid models are formed by combining different types of methods. In order to incorporate expert knowledge, interactive systems are also designed and realized. In this paper, we provide a review of the popular statistical and machine learning models, as well as the hybrid models and interactive systems based on the data-driven methods used for anomaly detection based on AIS data.

4.1 Introduction

Maritime transportation plays an important role in the development of the global economy, making maritime safety and security becoming an area of focus [1–4]. Maritime safety and security are monitored by the widely used surveillance and tracking systems. One of the typical and rapid developing systems is the space-born Automatic Identification System (AIS). The AIS transmitter is compulsive for all passenger ships and all other ships with more than 300 deadweight ton, and it is able to broadcast the static vessel data (including vessel IMO number, vessel name, vessel type, and vessel size, etc.) as well as kinematic vessel data (including vessel position, vessel speed, ship course, ship heading, etc.) [5]. The historical and live AIS records can be obtained from the Internet sources, including but not limited to Marina Traffic, VT explorer, HIS global, and AIShub. According to the review of Tu et al. on the AIS

R. Yan · S. Wang (✉)
Department of Logistics and Maritime Studies, Hong Kong Polytechnic University, Hung Hom, Hong Kong
e-mail: wangshuaian@gmail.com

R. Yan
e-mail: angel-ran.yan@connect.polyu.hk

© Springer Nature Singapore Pte Ltd. 2019
X. Qu et al. (eds.), *Smart Transportation Systems 2019*, Smart Innovation,
Systems and Technologies 149, https://doi.org/10.1007/978-981-13-8683-1_4

data quality [5], the commercial AIS data providers can provide accurate ship position information, and most of them can offer necessary vessel kinematic information. However, vessel kinematic information provided by the noncommercial AIS data sources is missed, and both the commercial and noncommercial data sources miss the static information of the vessels. Regarding the time resolution of AIS data, which refers to the time interval between two consecutive AIS message and is important for efficient data mining, it is unable to be guaranteed by the data providers.

Although first developed for avoiding ship collision accidents, the AIS data is widely used in the domain of maritime safety, including but not limited to real-time anomaly detection, ship route estimation, ship path planning, and ship and port performance evaluation. In addition, innovative and real-time maritime surveillance systems are also proposed by combining AIS data with visualization technologies.

Vessel anomaly detection is one of the most important application areas of AIS data. The coastal authorities use anomaly detection systems to identify the nefarious activities, such as piracy, human trafficking, drug smuggling, or abnormal ship sailing. In this paper, we present a review of the data-driven methods that are most widely used in vessel anomaly detection: the statistical methods and machine learning methods, and their combinations. In addition, the interactive systems based on the two methods are also introduced.

4.2   Literature Research Method and Review Structure

To access the related papers, we searched the database of Google Scholar, Web of Science, and Scopus by using the combination of "AIS" and "anomaly detection", or "AIS" and "abnormal behavior detection" as keywords. A total of 59 papers were found, including three review papers and 56 research papers. Then, papers using statistical and machine learning methods are chosen, with a total of 34 papers. Among the 34 selected paper, 12 papers adopt statistical methods, 10 papers adopt machine learning methods, 3 papers combine different methods, while 9 papers propose vessel anomaly detection systems. In the rest of the paper, we will make a detailed review of the related papers and give some possible research topics for future research.

4.3   Data-Driven Models for Vessel Anomaly Detection

Two types of data-driven methods are widely used in maritime vessel anomaly detection based on AIS data: statistical models and machine learning models [6]. Basically, two steps are composed in the data-driven methods. First, constructing a ship normal behavior model from historical and/or real-time data (i.e., from the training data). Second, matching newly observed vessel data (i.e., the training data) with the learned normalcy model. As the statistical models are constructed based on the assumption that in a stochastic process, normal cases occur with a higher probability than those abnormal cases, observations with a lower probability than the threshold in statistical inference process are considered as anomalous behaviors [7]. While in the machine learning models, testing vessels with properties that deviate from the learned normal model to some certain extent are claimed with abnormal behaviors [8]. By combining different types of methods, hybrid vessel anomaly detection models with better performance can be formed.

4.3.1 Statistical Models for Vessel Anomaly Detection

As a mathematical model based on probability distributions of random variables, statistical models are widely used in maritime vessel anomaly detection [6, 7, 8]. The main methods include but are not limited to Gaussian process, kernel density estimation, Gaussian mixture model, Bayesian networks, and hidden Markov chain.

One mainstream of research using statistical methods develops models based on Gaussian process. In 2011, Will et al. developed a model based on Gaussian process to demonstrate normal shipping behavior constructed from the AIS data. The model could detect the ship abnormal and unsafe behaviors based on the ship's speed and location [9]. Smith et al. combined Gaussian process and extreme value theory to detect ship abnormality, in which the extreme value distribution was endowed with vessel dynamic [10]. Base on this work, Smith et al. proposed another anomaly detection model combing Gaussian process, extreme value theory and divergence measurement. Distribution of vessel dynamic changes and data sample size changes were considered in the combined Gaussian process-extreme value theory (GP-EVT) model for the first time [11].

The kernel density estimation (KDE), which is a purely nonparametric statistical approach, is also used for ship anomaly detection. A simple and fast algorithm based on KDE for predicting the location and velocity of ships was proposed by Ristic et al., which could also be used for ship abnormal behavior detection [12]. On the contrary to KDE, the Gaussian mixture model (GMM), which is a parametric density estimation model, is also widely used. Laxhammar proposed two unsupervised normal vessel traffic pattern clustering models which were based on the GMM and used the Expectation-Maximization (EM) algorithm as the clustering algorithm. One primary model was based on the momentary velocities of the vessel in a two-dimensional plane, while the other was an extension of the primary model which incorporated the momentary position [13]. Some studies have compared the performance of KDE and GMM in ship abnormal behavior detection. Laxhammar et al. adopted a novel performance measure to evaluate and compare the vessel anomaly detection performance of the adaptive KDE and GMM. The results suggested that although KDE was superior in modeling normalcy of the ships, there was no significant difference in KDE and GMM when conducting anomaly detection [14]. Another comparison between the GMM and KDE was made by Anneken et al. By using an annotated dataset of real AIS data, the authors concluded that the false detections and the undiscovered anomalies were high for both algorithms [15]. To improve the detection accuracy, Brax et al. then introduced precise and imprecise anomaly detectors to extend the currently used anomaly detection approach by using Bayesian and credal combination operators [16].

As an easily understandable probabilistic graphical model, Bayesian network (BN) is also used in maritime anomaly detection. Lane et al. proposed a BN to detect the overall threat of the ships based on the identified five common anomalous ship behaviors [17]. In 2014, Mascaro et al. constructed dynamic and static BN models from real AIS data and incorporated information related to weather, time and ship dynamic motion in successive steps [18]. Apart from Bayesian networks, other methods based on Bayes' theorem are also chosen in anomaly detections. For example, Kowalska put forward a data-driven nonparametric Bayesian model based on Gaussian process. Active learning paradigm was adopted to reduce computational complexity [19].

In addition to the abovementioned statistical methods, models based on other forms of statistical approach are also proposed. For example, a computation method to detect abnormal vessel activities based on density mapping and hidden Markov model (HMM) was proposed by Tun et al. based on AIS data [20]. Smith et al. proposed a multi-class hierarchy framework for different class of ships based on conformal predictors. The model consisted of three levels of artificial classes: Global class for all normal AIS records, Type classes for each vessel type, and Local class for each vessel. As every vessel was treated individually, more abnormal ships could be detected [21].

4.3.2 Machine Learning Methods for Vessel Anomaly Detection

As a fast developing approach, machine learning methods are widely adopted in anomaly detection [22]. One typical type is unsupervised learning methods, which use unlabeled training data to construct the learning model by using clustering algorithms [23]. In contrast to unsupervised learning methods, supervised learning methods use training data with labels to construct the learning model [24].

One typical unsupervised approach used in vessel anomaly detection is the density-based spatial clustering of applications with noise (DBSCAN) data clustering algorithm, which was first proposed by Ester et al. [25]. When applied to maritime anomaly detection, Pallotta et al. proposed a methodology named Traffic Route Extraction and Anomaly Detection (TREAD) based on DBSCAN. The methodology was used to learn a statistical model from AIS data in an unsupervised way in order to turn the raw AIS data to information supporting decisions [26, 27]. Pallotta et al. further improved the TREAD methodology by using hierarchical reasoning. Based only on the positional information obtained from raw AIS data, the off-route vessels, which referred to the vessels that were not following an existing route, were first detected [28]. For those on-route vessels, ships with heading anomaly and speed anomaly were also identified. Liu et al. put forward an extended DBSCAN called DBSCANSD taking ship speed and direction as nonspatial attributes into consideration. Then, an algorithm to extract normal ship stopping areas was proposed to identify the ship moving and stopping areas [29]. Based on the clustering results of this work, Liu et al. extended the model into three division distances by considering longitude, latitude, speed, and direction to Improve detection accuracy [30]. As another extension of DBSCAN, Radon et al. proposed a three-layer framework called MADCV combining weather information as "context information" to filter false alarms. Empirical studies using AIS data suggested that this approach could adapt to new contextual information [31].

Apart from DBSCAN, other unsupervised clustering methods were also included in the studies on anomaly detection. For example, a framework named MT-MAD, which was short for Maritime Trajectory Modeling and Anomaly Detection was proposed by Lei. As the name suggested, the model contained two parts: one for unsupervised maritime trajectory modeling and the other for anomaly detection. Experimental results on real AIS data showed that the proposed framework was effective in detecting maritime anomalous movement behaviors [32].

Other forms of unsupervised machine learning methods were also proposed based on AIS data. Vespe et al. proposed an unsupervised approach to learn ship motion patterns by creating ship waypoint graph, which contained vessel objects, turning point, ports, and offshore platforms, entry or exit point, sea lane, and route [33]. Guillarme and Lerouvreur proposed a probabilistic based normalcy model of vessel dynamics learned in an unsupervised way. The model had two levels and mainly contained three parts: trajectory partitioning, clustering and path modeling. The two-scale model could be used for anomaly detection, AIS consistency analysis, and ship path prediction [34].

4.3.3   Hybrid Models for Vessel Anomaly Detection

Hybrid models for vessel abnormal behavior detection are formulated by combining different forms of methods or combining the numerical models with expert knowledge to improve the prediction accuracy. Nevertheless, the hybrid models are not easy to be implemented due to its complicated format. By counteracting the drawbacks of different models, several studies have developed high-performance models for vessel anomaly detection based on AIS data. De Vries and van Someren proposed a machine learning network to perform tasks including ship trajectories clustering, classification and outlier detection. There were three steps in the approach. First, a piecewise linear segmentation method was adopted to compress trajectories. Second, a similarity-based approach based on kernel methods was applied for detection tasks of clustering, classifying an outlier. Finally, geographical domain knowledge was integrated with the framework [35]. Smith et al. described conformal prediction techniques for maritime anomaly detection. First, an unsupervised conformal anomaly detection framework was proposed under the criteria of the average p-

value. Then, two supervised non-conformity measures were adopted: one was based on KDE, and the other was based on k-nearest neighbors (k-NN) algorithm as a comparison. After that, AIS data was used to test the performance of the non-conformity models with average p-value as a criterion. The results indicated that KDE performed better than k-NN as a non-conformity measure with the data set [36]. Wang et al. presented a two-level approach for vessel abnormal routes detection. First, an unsupervised model named DBSCANSD which was proposed by Liu et al. [29] was applied for data points pre-clustering. Then, combined with the expert knowledge, optimal labeling results of the data points (normal or abnormal) were generated. Second, a supervised learning model was trained by generated labeled data to detect the anomaly of new-coming vessels [37].

4.4 Interactive Systems for Vessel Anomaly Detection

As the detection of ship anomalous behaviors is a complicated problem, especially when confronting the rare and ambiguous situations, support and guidance from the domain experts are beneficial to the detection process [38]. By applying interactive and visualization technologies, expert judgment and decision can be comprised in the vessel anomaly detection systems.

The basis of one typical ship anomaly detection system was laid by a study conducted by Rhodes et al., in which a learning model comprising an unsupervised clustering algorithm and a supervised labeling and mapping algorithm were proposed. Both the vessel routine behaviors and abnormal behaviors were learned by these cognitively inspired algorithms [39]. To improve the performance of the proposed system, neuro-biologically inspired algorithms were added for better adapting to the evolving situations [40]. Then, an improved neuro-biologically inspired algorithm consisting of real-time tracking information and with the ability of learning motion patterns were proposed by Rhodes et al. Weights were added to the learned model, which enabled better ship location prediction performance [41]. In order to combine user knowledge in the anomaly detection system, an interactive approach to combine expert knowledge and normal behavior models was proposed by Riveiro et al. The approach was implemented on an anomaly detection prototype called VISAD [38]. Then, Riveiro et al. gave an overview introduction of VISAD, in which expert knowledge was inserted and the detection procedure was transparent to the users [42]. Another surveillance system called SeeCoast was illustrated by Rhodes et al. In SeeCoast, vessel track classification information was generated based on the video streams provided by USCG cameras. Then, a coherent track picture was generated according to the AIS data and surface surveillance radars. Based on the track picture, the unsafe, illegal, and threatening vessel activities could be identified using machine learning methods [43]. Other similar systems, such as the satellite-extended-vessel Traffic Service (SEV) system [44] and distributed multi-hypothesis tracking (DMHT) technology-based trackers [45] were also designed, introduced and analyzed in the literature.

4.5   Future Research Opportunities

The access to shipping AIS data makes it possible to apply statistical and machine learning methods to maritime anomaly detection. Undoubtedly, the application of AIS data to vessel abnormal behavior detection will receive more attention in the future. In this section, we will provide some future research topics on maritime anomaly detection based on AIS data.

First, most of the abovementioned papers aim to focus on offline anomalous detection, i.e., using the historical AIS data to construct the detection models and test the performance of the detection models. As online anomalous vessel behavior detection is not allowed, implementation of the proposed frameworks and models will be hindered. Therefore, one possible research area is developing real-time detection algorithms and systems to identify online vessel abnormal behaviors. Second, most of the models designed for vessel anomaly detection only consider the vessel kinematic attributes, such as the vessel speed over ground, vessel course over ground and vessel location information. However, vessel static factors, including vessel type, vessel size, vessel flag, vessel company and non-vessel factors, including the regional traffic density, local weather conditions, regional policies and regulations will also have an impact on the vessel behaviors. Taking those factors into consideration when constructing the normal behavior models may help improve detection accuracy. Third, the performance of the detection ability of the models is largely dependent on the quality of AIS data. However, there are some inaccuracies and errors occurring in the data, especially the manually entered data such as ship's length, IMO number, name, call sign, etc. Moreover, not all ships are equipped with or properly operate the AIS system, so not all ships can be well detected. Thus, combining different sources of data, including but not limited to Long-Range Identification and Tracking of ships (LRIT), Synthetic Aperture Radar (SAR), sensors like sonar and thermal infrared, as well as identifying and filtering incorrect AIS data [45], can help refine the input data, thus improving the performance of the detection models.

References

1. Weng, J., Yang, D., Qian, T., Huang, Z.: Combining zero-inflated negative binomial regression with MLRT techniques: an approach to evaluating shipping accident casualties. Ocean Eng. **166**, 135–144 (2018)
2. Weng, J., Meng, Q., Qu, X.: Vessel collision frequency estimation in the Singapore Strait. J. Navig. **65**(2), 207–221 (2012)
3. Li, S., Meng, Q., Qu, X.: An overview of maritime waterway quantitative risk assessment models. Risk Anal.: Int. J. **32**(3), 496–512 (2012)
4. Qu, X., Meng, Q., Li, S.: Ship collision risk assessment for the Singapore Strait. Accid. Anal. Prev. **43**(6), 2030–2036 (2011)
5. Tu, E., Zhang, G., Rachmawati, L., Rajabally, E., Huang, G.B.: Exploiting AIS data for intelligent maritime navigation: a comprehensive survey from data to methodology. Proc. IEEE Trans. Intell. Transp. Syst. **19**(5), 1559–1582 (2018)

6. Smith, M.: Anomaly detection in vessel track data. (Thesis submitted for the degree of Master of Science (Research)). http://www.robots.ox.ac.uk/~parg/pubs/theses/markSmithMScThesis2014.pdf. Accessed Dec 2018
7. Anscombe, F.J.: Rejection of outliers. Technometrics **2**(2), 123–146 (1960)
8. Sidibé, A., Shu, G.: Study of automatic anomalous behavior detection techniques for maritime vessels. J. Navig. **70**(4), 847–858 (2017)
9. Will, J., Peel, L., Claxton, C.: Fast maritime anomaly detection using kd-tree gaussian processes. In: Proceedings of IMA Maths in Defence Conference (2011)
10. Smith, M., Reece, S., Roberts, S., Rezek, I.: Online maritime abnormality detection using gaussian processes and extreme value theory. In: Proceedings of IEEE 12th International Conference on Data Mining, pp. 645–654 (2012)
11. Smith, M., Reece, S., Roberts, S., Psorakis, I., Rezek, I.: Maritime abnormality detection using Gaussian processes. Knowl. Inf. Syst. **38**(3), 717–741 (2014)
12. Ristic, B., La Scala, B.F., Morelande, M.R., Gordon, N.J.: Statistical analysis of motion patterns in AIS data: Anomaly detection and motion prediction, FUSION, pp. 1–7 (2008)
13. Laxhammar, R.: Anomaly detection for sea surveillance. In: Proceedings of 2008 11th IEEE International Conference on Information Fusion, pp. 1–8 (2008)
14. Laxhammar, R., Falkman, G., Sviestins, E.: Anomaly detection in sea traffic-a comparison of the gaussian mixture model and the kernel density estimator. In: Proceedings of IEEE 2009 12th International Conference on Information Fusion, pp. 756–763 (2009)
15. Anneken, M., Fischer, Y., Beyerer, J.: Evaluation and comparison of anomaly detection algorithms in annotated datasets from the maritime domain. In: Proceedings of IEEE SAI Intelligent Systems Conference, pp. 169–178 (2015)
16. Brax, C., Karlsson, A., Andler, S.F., Johansson, R., Niklasson, L.: Evaluating precise and imprecise state-based anomaly detectors for maritime surveillance. In: Proceedings of IEEE 2010 13th International Conference on Information Fusion, pp. 1–8 (2010)
17. Lane, R.O., Nevell, D.A., Hayward, S.D., Beaney, T.W.: Maritime anomaly detection and threat assessment. In: Proceedings of IEEE 2010 13th International Conference on Information Fusion, pp. 1–8 (2010)
18. Mascaro, S., Nicholso, A.E., Korb, K.B.: Anomaly detection in vessel tracks using Bayesian networks. Int. J. Approx. Reason. **55**(1), 84–98 (2014)
19. Kowalska, K., Peel, L.: Maritime anomaly detection using Gaussian process active learning. In: Proceedings of IEEE 2012 15th International Conference on Information Fusion, pp. 1164–1171 (2012)
20. Tun, M.H., Chambers, G.S., Tan, T., Ly, T.: Maritime port intelligence using AIS data. In: Recent Advances in Security Technology, pp. 33–43 (2007)
21. Smith, J., Nouretdinov, I., Craddock, R., Offer, C., Gammerman, A.: Conformal anomaly detection of trajectories with a multi-class hierarchy. In: International Symposium on Statistical Learning and Data Sciences, pp. 281–290 (2015)
22. Zhou, M., Qu, X., Li, X.: A recurrent neural network based microscopic car following model to predict traffic oscillation. Transp. Res. Part C **84**, 245–264 (2017)
23. Hastie, T., Tibshirani, R., Friedman, J.: Unsupervised Learning, pp. 485–585. Springer Press, New York (2009)
24. Jordan, M.I.: Supervised learning and systems with excess degrees of freedom (1988)
25. Ester, M., Kriegel, H.P., Sander, J., Xu, X.: A density-based algorithm for discovering clusters in large spatial databases with noise. ACM Knowl. Discov. Data Min. **96**(34), 226–231 (1996)
26. Pallotta, G., Vespe, M., Bryan, K.: Vessel pattern knowledge discovery from AIS data: A framework for anomaly detection and route prediction. Entropy **15**(6), 2218–2245 (2013)
27. Pallotta, G., Vespe, M., Bryan, K.: Traffic knowledge discovery from AIS data. In: Proceedings of IEEE 2013 16th International Conference on Information Fusion (FUSION), pp. 1996–2003 (2013)
28. Pallotta, G., Jousselme, A.L.: Data-driven detection and context-based classification of maritime anomalies. In: Proceedings of IEEE 2015 18th International Conference on Information Fusion (Fusion), pp. 1152–1159 (2015)

29. Liu, B., de Souza, E.N., Matwin, S., Sydow, M.: Knowledge-based clustering of ship trajectories using density-based approach. In: Proceedings of IEEE 2014 International Conference on Big Data, pp. 603–608 (2014)
30. Liu, B., de Souza, E.N., Hilliard, C., Matwin, S.: Ship movement anomaly detection using specialized distance measures. In: Proceedings of IEEE 2015 18th International Conference on Information Fusion (Fusion), pp. 1113–1120 (2015)
31. Radon, A.N., Wang, K., Glässer, U., Wehn, H.: Westwell-Roper, contextual verification for false alarm reduction in maritime anomaly detection. In: Proceedings of IEEE 2015 International Conference on Big Data, pp. 1123–1133 (2015)
32. Lei, P.R.: A framework for anomaly detection in maritime trajectory behavior. Knowl. Inf. Syst. **47**(1), 189–214 (2016)
33. Vespe, M., Visentini, I., Bryan, K., Braca, P.: Unsupervised learning of maritime traffic patterns for anomaly detection. In: Proceedings of 9th IET Data Fusion and Target Tracking Conference, pp. 1–5 (2012)
34. Le Guillarme, N., Lerouvreur, X.: Unsupervised extraction of knowledge from S-AIS data for maritime situational awareness. In: Proceedings of 2013 16th International Conference on Information Fusion (FUSION), pp. 2025–2032 (2013)
35. De Vries, G.K.D., Van Someren, M.: Machine learning for vessel trajectories using compression, alignments and domain knowledge. Expert Syst. Appl. **39**(18), 13426–13439 (2012)
36. Smith, J., Nouretdinov, I., Craddock, R., Offer, C., Gammerman, A.: Anomaly detection of trajectories with kernel density estimation by conformal prediction. In: Proceedings of IFIP International Conference on Artificial Intelligence Applications and Innovations, pp. 271–280 (2014)
37. Wang, X., Liu, X., Liu, B., de Souza, E.N., Matwin, S.: Vessel route anomaly detection with Hadoop MapReduce. In: Proceedings of IEEE International Conference on Big Data, pp. 25–30 (2014)
38. Riveiro, M., Falkman, G.: Interactive visualization of normal behavioral models and expert rules for maritime anomaly detection. In: Proceedings of 2009 Sixth International Conference on Computer Graphics, Imaging and Visualization, pp. 459–466 (2009)
39. Rhodes, B.J., Bomberger, N.A., Seibert, M., Waxman, A.M.: Maritime situation monitoring and awareness using learning mechanisms. In: Proceedings of IEEE Military Communications Conference, pp. 646–652 (2005)
40. Bomberger, N.A., Rhodes, B.J., Seibert, M., Waxman, A.M.: Associative learning of vessel motion patterns for maritime situation awareness. In: Proceedings of 2006 9th International Conference on Information Fusion, pp. 1–8 (2006)
41. Rhodes, B.J., Bomberger, N.A., Zandipour, M.: Probabilistic associative learning of vessel motion patterns at multiple spatial scales for maritime situation awareness. In: Proceedings of IEEE 2007 10th International Conference on Information Fusion, pp. 1–8 (2007)
42. Riveiro, M., Falkman, G., Ziemke, T., Warston, H.: VISAD: an interactive and visual analytical tool for the detection of behavioral anomalies in maritime traffic data. Vis. Anal. Homel. Def. Secur. **7346** (2009)
43. Rhodes, B.J., Bomberger, N.A., Freyman, T.M., Kreamer, W., Kirschner, L., Adam, C.L., Mungovan, I.W., Stauffer, C., Stolzar, L., Waxman, A.M., Seibert, M.: SeeCoast: persistent surveillance and automated scene understanding for ports and coastal area. Def. Transform. Net-Centric Syst. **6578** (2007)
44. Vespe, M., Sciotti, M., Burro, F., Battistello, G., Sorge, S.: Maritime multi-sensor data association based on geographic and navigational knowledge. In: Proceedings of IEEE 2008 Radar Conference, pp. 1–6 (2008)
45. Carthel, C., Coraluppi, S., Grignan, P.: Multisensor tracking and fusion for maritime surveillance. In: Proceedings of IEEE 2007 10th International Conference on Information Fusion, pp. 1–6 (2007)

Chapter 5
Estimation Method of Saturation Flow Rate for Shared Left-Turn Lane at the Signalized Intersection, Part I: Methodology

Ruru Tang, Yunhao Wang, Yiming Bie and Zhiqiang Fang

Abstract Saturation flow rate (SFR) estimation of approaching lanes is an essential task to evaluate the signal timing and level of service of signalized intersections. There are few existing studies on SFR estimation of the shared left-turn lane. According to relative differences of directions between the front and behind vehicles, this paper divides the headways of shared left-turn lane during green times into four categories. For each category, a method is raised to identify whether the headways are saturated based on the KPSS test. The paper first provides a method to eliminate outliers considering the factors such as distracted driving and stochastic flow. The estimated value of SFR should be 3600 divided by saturation headway, and the estimated value of saturation headway is calculated as the weighted average value of four categories.

5.1 Introduction

Saturation flow rate (SFR) estimation is an essential task in the operation stage of signalized intersections. SFR is a fundamental input parameter for capacity calculation, signal timing design and level of service evaluation [1, 2]. Traditionally, SFR is defined as the maximum number of vehicles in a period of time (commonly in one hour green) that can pass through a given stop line of the approaching lanes

R. Tang
School of Transportation Science and Engineering, Harbin Institute of Technology, Harbin 150090, China
e-mail: 2603332567@qq.com

Y. Wang
School of Economics, Northeast Normal University, Changchun 130117, China
e-mail: wangyh533@foxmail.com

Y. Bie (✉)
School of Transportation, Jilin University, Changchun 130022, China
e-mail: yimingbie@126.com

Z. Fang
Faculty of Science, University of Melbourne, 3010 Melbourne, VIC, Australia
e-mail: zhiqfang@foxmail.com

© Springer Nature Singapore Pte Ltd. 2019
X. Qu et al. (eds.), *Smart Transportation Systems 2019*, Smart Innovation, Systems and Technologies 149, https://doi.org/10.1007/978-981-13-8683-1_5

[3]. However, this estimation task becomes difficult to perform satisfactorily when shared lanes are present.

There are continuous researches on the estimation method of SFR. The main aim of those researches is to recognize steady traffic flow and obtain saturation headway, and thus calculate SFR. Some researchers thought saturation headway was inverse proportional to the saturation flow rate that was defined as the maximum number of vehicles that could be serviced in an hour of green [4–6]. Niittymaki and Pursula defined that departure headway was the elapsed time between every two consecutive vehicles, when they cross the stop line one by one after the light turning green [7]. Hossain introduced a new microscopic simulation technique to estimate the saturation flow from the influencing variables like road width, turning proportion, and percentage of heavy and non-motorized vehicles [8]. Jin et al. found that the distributions of the departure headways at each position in the discharging queue were revealed to approximately follow a certain log-normal distribution (except the first one) and the corresponding mean values leveled out gradually [9]. Yang et al. provided a dynamic extraction method of saturation flow rate by using the induction coil detector. The average saturation headways of history and current cycle were dealt with exponential smoothing method [10]. Hao and Ma presented that Shifted Log-Normal was more suitable to capture the distribution of steady-flow headways which were approximately the headways at positions between the fifth vehicle and the last stopping vehicle in the queue and the mean could be considered as practical saturation headway which was used for actual capacity analysis in practice [11]. Wang et al. proposed an automatic estimation method for the SFR based on video detector data and the Dickey–Fuller test was used to verify whether the headways in the time series were saturation headways. Meanwhile, they employed an iterative method using quantiles to filter out abnormal data. The SFR was finally calculated using the average value of saturation headways [12].

As mentioned above, although there are many research outcomes on SFR estimation, it is hard to apply them directly to the SFR estimation of a shared left-turn lane. The reason is headway of shared left-turn lane contains four categories which are headway between two going through vehicles, headway between going through vehicle and left-turning vehicle, headway between two left-turning vehicles and headway between left-turning vehicle and going through vehicles. There are large differences among the headways of different categories even if undersaturation condition. That is to say, it is hard to estimate SFR under this circumstance and existing studies cannot be applied to shared left-turn lane.

The objective of this study is to establish an SFR estimation method for a shared left-turn lane. In this paper, we first divide the headway of shared left-turn lane into four categories and then a stationarity test method is proposed to recognize the saturation headway of each category. Finally, the saturation headway of a shared left-turn lane is calculated as the weighted average value of four categories' saturation headway. Therefore, the SFR of a shared left-turn lane can be obtained.

5.2   The Category of Headway

Headway is considered as the time intervals between two successive vehicles passing stop line or pre-set reference line at the same intersections during the green time. To calculate SFR, the headway under saturation movement state, that is saturation headway, is required. The relationship between SFR S_μ and saturation headway h_μ can be expressed as Eq. (5.1) follows:

$$S_\mu = 3600/h_\mu \qquad (5.1)$$

In the equation:

S_μ (pcu/h):   SFR of single lane.
h_μ (s):   average headway during saturation movement state

According to Eq. (5.1), it is essential to have saturation headway while calculating SFR. During green time, vehicles at shared left-turn lane pass the stop line and form a traffic flow. Even though during saturation state, the headways vary dramatically if the one headway changes due to the different directions involved in front and behind vehicles. Therefore, to estimate SFR, the flow directions of involved front and behind vehicles have to be considered. This paper divides headways into four categories as follows:

(i)   headway between two going through vehicles.
(ii) headway between going through vehicle and left-turning vehicle.
(iii) headway between left-turning vehicle and going through vehicles.
(iv) headway between two left-turning vehicles.

Obviously, headway is relatively smaller for category (i) and larger for category (iv). The speed of through vehicles is larger than that of left-turn vehicles, because left-turn vehicles need to decelerate to complete the turning at the intersection. Furthermore, according to the survey data at the practical signalized intersection, the headway of category (i) is also smaller than category (iv). So, there are great differences in headways even if one shared left-turn lane is at saturation state. Under this circumstance, it is difficult to determine whether the traffic flow is at saturation or not by setting a constant threshold value. In this case, the paper provides a method for each category to identify saturation headway and then calculates saturation headway of shared left-turn lane by the weighted average method.

5.3   Saturation Headway Recognition

This paper adopts a stationarity test method proposed by Kwiatkowski and Phillips et al. in 1992, usually called the KPSS test, to recognize saturation headways [13]. The basic idea of the KPSS test is shown as follows:

Let y_t, $t = 1, 2, \ldots, T$, be the observed series for which we wish to test stationarity. Kwiatkowski et al. assumed that they can decompose the series into the sum of a deterministic trend, a random walk, and a stationary error:

$$y_t = \xi t + r_t + \varepsilon_t \tag{5.2}$$

Here r_t is a random walk:

$$r_t = r_{t-1} + u_t \tag{5.3}$$

Where the u_t are iid $(0, \sigma_u)$. The initial value r_0 is treated as fixed and serves the role of an intercept. The stationarity hypothesis is simply $\sigma_u = 0$. Since ε_t is assumed to be stationary, under the null hypothesis y_t is trend stationary. They also consider the special case of the model (5.2) in which they set $\xi = 0$, in which case under the null hypothesis y_t is stationary around a level (r_0) rather than around a trend.

During green times, if the traffic flow at the intersection is undersaturated operation state, discharging vehicles will follow the vehicle in front of them closely. Their headways will have a small difference, which approach to the saturation headway, so they should fluctuate around saturation headway. Due to the huge difference between the four categories, they should be studied separately. Take a certain category as an example, pick up all headways belong to the category to form a time series $\{h_m\}_{m=1}^{M}$, M is the number of samples of the series. Then the mth headway sample h_m should fluctuate irregularly around the actual saturation headway μ. That is

$$h_m = \mu + \varepsilon_m \tag{5.4}$$

In the equation

μ (s):   actual saturation headways
ε_m: random error term of mth headway

$\{\varepsilon_m\}_{m=1}^{M}$, the time series formed by ε_m obeying normal distribution with a mean of 0 and variance of σ_ε^2, each random error term in the time series is independently and identically distributed, recorded as $\varepsilon_m \sim i.i.d\ N(0, \sigma_\varepsilon^2)$. Because μ is the unknown variable required to be estimated. The actual fluctuation around μ is also unknown. There for σ_ε^2 is unknown.

The paper determines whether actual headway time series $\{h_m\}_{m=1}^{M}$ is saturation headway time series using the KPSS test. At the very beginning, constitute auxiliary regression equations as shown below:

$$h_m = \alpha_m + e_m \tag{5.5a}$$

$$\alpha_m = \alpha_{m-1} + \eta_m \tag{5.5b}$$

In the equations

α_m: the estimation value of h_m at mth time slot.
$e_m \sim i.i.d\ N(0, \sigma_e^2)$:   random error term.
$\eta_m \sim i.i.d\ N(0, \sigma_\eta^2)$:   random error term

Based on the KPSS test, the first step is to determine whether the variance of random error term η_m is zero or not. Which means, under null hypothesis, $\sigma_\eta^2 = 0$, if the null hypothesis is true, calculate means and variances for both sides of formula (5.5b):

$$E(\alpha_m) = E(\alpha_{m-1}) + E(\eta_m) = E(\alpha_{m-1}) \qquad (5.6a)$$

$$Var(\alpha_m) = Var(\alpha_{m-1}) + Var(\eta_m) = Var(\alpha_{m-1}) \qquad (5.6b)$$

It can be seen from Eq. (5.6a) that if null hypothesis is true, the estimation value of mth headway α_m and m – 1th headway α_{m-1} have the same exception and variance. Meanwhile, they should equal to a constant: $\alpha_m = \alpha_{m-1} = \alpha$. Now, Eq. (5.5a) could be simplified as

$$H_m = \alpha + e_m \qquad (5.7)$$

It can be seen that, Eq. (5.7) and (5.4) have the same structure. That is to say, when the null hypothesis is true, time series $\{h_m\}_{m=1}^{M}$ is the time series of saturation headways.

To test whether the variance of the random error term η_m is 0 or not, KPSS test firstly use ordinary lest square method (OLS method) to estimate the residuals of Eq. (5.7), which is

$$\hat{e}_m = h_m - \frac{1}{M} \sum_{m=1}^{M} h_m \qquad (5.8)$$

Second, establish the series of residual $\hat{e}_m$:

$$S_m = \sum_{i=1}^{m} \hat{e}_i \qquad (5.9)$$

Then, KPSS build a function $S^2(l)$ with parameter l involved:

$$S^2(l) = M^{-2} \sum_{m=1}^{M} \hat{e}_m^2 + 2M^{-2} \sum_{s=1}^{l} w(s, l) \sum_{m=s+1}^{M} \hat{e}_m \hat{e}_{m-s} \qquad (5.10)$$

Where $w(s, l) = 1 - s/(l + 1)$.
Finally, the statistical value can be expressed as

Table 5.1 The critical value for τ corresponding to different upper quantiles

Upper quantile	0.1	0.05	0.025	0.01
Critical value	0.347	0.463	0.574	0.739

$$\tau = M^{-2} \sum_{m=1}^{M} S_m^2 / S^2(l) \tag{5.11}$$

In this function, l is an optional parameter which is suggested as $l = 12(M/100)^{1/4}$ by KPSS test.

KPSS test is a kind of one-sided test, the critical value of parameter τ is shown in Table 5.1. The 10% upper quantile of τ is 0.347, which means at a significant level of 0.1, if τ is less than 0.347, the null hypothesis cannot be rejected. This headways series can be regarded as headways series under saturation state.

5.4 Elimination of Abnormal Headways

Section 5.3 provided the method of determining whether headways time series is saturation headways or not. In this section, a method to eliminate outliers will be introduced.

Actual traffic flow data are required in SFR estimation. Assuming that the data start to be collected at t_a and end at t_b. We record the time of all vehicles passing stop line to calculate the original headway time series $\{y_{0,k}\}_{k=1}^{K_1}$ during time period $[t_a, t_b]$. The number of samples in the series is K_1, kth sample is $y_{0,k}$ ($1 \leq k \leq K_1$), Because of distracted driving or stochastic flow at the end of green times, headways become larger and not suitable for SFR estimation. Therefore, an elimination of those abnormal data are required for series $\{y_{0,k}\}_{k=1}^{K_1}$.

There are two kinds of abnormal headways in series $\{y_{0,k}\}_{k=1}^{K_1}$: (i) During the green time, the first several vehicles are still in the process of acceleration, the headway is rather large; (ii) During the green time, because of distracted driving or stochastic flow, headways will become large also.

According to this, an iteration step has been designed to eliminate the abnormal data from $\{y_{0,k}\}_{k=1}^{K_1}$ as shown below. After those steps, the time series containing saturation headways could pass the KPSS test as mentioned in Sect. 5.3.

Step 1: set the initial value of i to 1, that is $i = 1$.

Step 2: KPSS test on series $\{y_{i,k}\}_{k=1}^{K_i}$, if the null hypothesis cannot be rejected, turn to step 5, then $\{y_{i,k}\}_{k=1}^{K_i}$ can be regarded as a series of saturation headways. Else, turn to step 3.

Step 3: Calculate the mean value $\frac{1}{K_i} \sum_{k=1}^{K_i} y_{i,k}$ of headways in series $\{y_{i,k}\}_{k=1}^{K_i}$, then calculate the distance between $y_{i,k}$ ($1 \leq k \leq K_i$) and the mean value.

Eliminate the datum with the largest distance to get $\{y_{i+1,k}\}_{k=1}^{K_{i+1}}$ including K_{i+1} samples.

Step 4: $i = i + 1$, turn to Step 2.

Step 5: End of iteration.

In Step 3, the distance between $y_{i,k}$ and the mean value is $\left| y_{i,k} - \frac{1}{K_i} \sum_{k=1}^{K_i} y_{i,k} \right|$. The result after last cycle is $\{y_{n,k}\}_{k=1}^{K_n}$. The series contains K_n samples. It can be seen from the iteration steps that after each cycle, there will be one sample less, which means the number of samples K_n in $\{y_{n,k}\}_{k=1}^{K_n}$ is $K_1 - (n-1)$

5.5 SFR Calculation Method

As it is mentioned in Sect. 5.4, because the headways series $\{y_{n,k}\}_{k=1}^{K_n}$ of a certain category has passed the KPSS test, it is a series of saturation headways. According to Eq. (5.4) in Sect. 5.3, saturation headways obey normal distribution. That is to say, $\{y_{n,k}\}_{k=1}^{K_n}$ should obey normal distribution. Then the mean, median, and mode of $\{y_{n,k}\}_{k=1}^{K_n}$ should be roughly equal.

For convenient calculation, the paper use mean value to estimate the saturation headway μ, the estimation value is recoded as $\hat{\mu}$:

$$\hat{\mu} = \frac{1}{K_n} \sum_{k=1}^{K_n} y_{n,k} \tag{5.12}$$

According to the method mentioned in Sect. 5.4, the headway time series of shared left-turn lane can be divided into four categories, recorded as $\{h_{1,m_1}\}_{m_1=1}^{M_1}$, $\{h_{2,m_2}\}_{m_2=1}^{M_2}$, $\{h_{3,m_3}\}_{m_3=1}^{M_3}$, $\{h_{4,m_4}\}_{m_4=1}^{M_4}$, respectively, with M_1, M_2, M_3, M_4 samples in the series. The mean headways of kth category $\bar{h}_k$ is:

$$\bar{h}_k = \frac{1}{M_k} \sum_{m=1}^{M_k} h_{k,m}, \ k = 1, 2, 3, 4 \tag{5.13}$$

Meanwhile, the weighted headways of the shared left-turn lane, expressed as $\hat{h}_\mu$ can be estimated as

$$\hat{h}_\mu = \frac{1}{M_0} \sum_{k=1}^{4} M_k \bar{h}_k \tag{5.14}$$

In the equation

$$M_0 = \sum_{k=1}^{4} M_k: \quad \text{The total number of samples in four different categories.}$$

The estimation SFR $\hat{S}_\mu$, equal to $3600/h_\mu$, can be calculated with the known of weighted saturation headways. The reason for using the weighted average value of saturation headways in Eq. (5.14) is $\hat{h}_\mu$ related to the number of samples of different categories. When category (i) has more samples, the value $\hat{h}_\mu$ will become smaller and vice versa.

5.6 Conclusion

In this study, we divide the headways of shared left-turn lane into four categories and then use the KPSS test to generate a method to identify saturation headways and eliminate outliers. The weighted average of headways is considered as saturation headway of the shared left-lane lane. The premise of SFR estimation is to have saturation headways and the premise of getting saturation headways is the identification of saturation flows. Headway is an essential indicator of saturation flow. Based on KPSS test, the paper generated the identification methods for four categories. That is the key to SFR estimation for a shared left-turn lane. The impacts of weather condition, number of approaching lanes and buses will be examined in the future [14–17].

Acknowledgements This study is supported by the National Natural Science Foundation of China (NO. 71771062 & 71771050).

References

1. Qu, X., Zhang, J., Wang, S.: On the Stochastic Fundamental Diagram for Freeway Traffic: Model Development, Analytical Properties, Validation, and Extensive Applications. Transp. Res. Part B: Methodological. **104**, 256–271 (2017)
2. Zhou, M., Qu, X., Li, X.: A recurrent neural network based microscopic car following model to predict traffic oscillation. Transp. Res. Part C: Emerg. Technologies. **84**, 245–264 (2017)
3. Highway Capacity Manual 2000. Transportation Research Board, Washington, USA (2000)
4. Li, X., Medal, H., Qu, X.: Connected Heterogeneous Infrastructure Location Design under Additive Service Utilities. Transp. Res. Part B: Methodol. **120**, 99–124 (2019)
5. Stokes, R.W.: Comparison of saturation flow rates at signalized intersections. ITE J. **58**(11), 15–20 (1988)
6. Bonneson, J.A.: Modeling queued driver behavior at signalized junctions. Transp. Res. Rec. **1365**, 99–107 (1992)
7. Niittymaki, J., Pursula, M.: Saturation flow at signal-group-controlled traffic signals. Transp. Res. Rec.: J. Transp. Res. Board **1572**, 24–32 (1996)

8. Hossain, M.: Estimation of saturation flow at signalised intersections of developing cities: a micro-simulation modelling approach. Transp. Res. Part A: Policy Pract. **35**(2), 123–141 (2001)
9. Jin, X., Zhang, Y., Wang, F., Li, L., Yao, D., Sun, Y., Wei, Z.: Departure headways at signalized intersections: A log-normal distribution model approach. Transp. Res. Part C Emerg. Technol. **17**(3), 318–327 (2009)
10. Yang, D.Y., Luo, J.L., Liu, C., Duan, Z.Y.: Dynamic extraction method of saturation flow rate at signalized intersection. J. Traffic & Transp. Eng. **13**(1), 98–103 (2013)
11. Hao, H.M., Ma, W.J.: Revisiting distribution model of departure headways at signalised intersections. Transp. B: Transp. Dyn. **5**(1), 1–14 (2017)
12. Wang, L.H., Wang, Y.H., Bie, Y.M.: automatic estimation method for intersection saturation flow rate based on video detector data. J. Adv. Transp. **2018** (2018). Article ID. 8353084
13. Kwiatkowski, D., Phillips, P.C.B., Schmidt, P., Shin, Y.C.: Testing the null hypothesis of stationarity against the alternative of a unit root: How sure are we that economic time series have a unit root? J. Econ. **54**(1–3), 159–178 (1992)
14. Bie, Y., Liu, Z., Ma, D., Wang, D.: Calibration of platoon dispersion parameter considering the impact of the number of lanes. J. Transp. Eng. **139**(2), 200–207 (2013)
15. Wang, Y., Bie, Y., An, Q.: Impacts of winter weather on bus travel time in cold regions: case study of harbin, China. J. Transp. Eng. Part A: Syst. **144**(11) (2018). Article number: 05018001
16. Wang, Y., Bie, Y., Zhang, L.: Joint optimization for the locations of time control points and corresponding slack times for a bus route. KSCE J. Civ. Eng. **23**(1), 411–419 (2019)
17. Bie, Y., Gong, X., Liu, Z.: Time of day intervals partition for bus schedule using GPS data. Transp. Res. Part C: Emerg. Technol. **60**, 443–456 (2015)

Chapter 6
Estimation Method of Saturation Flow Rate for Shared Left-Turn Lane at Signalized Intersection, Part II: Case Study

Ruru Tang, Yunhao Wang, Yiming Bie and Zhiqiang Fang

Abstract This paper is a case study to estimate the Saturation Flow Rate (SFR) on a practical shared left-turn lane during peak hour and off-peak hour using KPSS test method and Highway Capacity Manual (HCM) method. The results declare that the value of SFR estimated by HCM is less than the one estimated by KPSS test method during all time periods. The difference is 4.1% during off-peak hour while 1.2% during peak hour. However, this slight discrepancy could cause remarkable difference in cycle length and average vehicle delay. For example, compared with KPSS test method, in off-peak hour the cycle length increases by 19.2% while the average vehicle delay increases by 14.0% when using HCM method.

6.1 Introduction

Considering the importance of the Saturation Flow Rate (SFR) for planning, design, and operation at the signalized intersection, many countries established calculation models for SFR [1–4]. Such as America, England, and Canada, among which the Highway Capacity Manual (HCM) in America is widely used [5]. For shared left-turn lanes which have exclusive signal phase, when the roads are newly built, HCM provided an adjustment method, which employs the product of the SFR of through

R. Tang
School of Transportation Science and Engineering, Harbin Institute of Technology, Harbin 150090, China
e-mail: 2603332567@qq.com

Y. Wang
School of Economics, Northeast Normal University, Changchun 130117, China
e-mail: wangyh533@foxmail.com

Y. Bie (✉)
School of Transportation, Jilin University, Changchun 130022, China
e-mail: yimingbie@126.com

Z. Fang
Faculty of Science, University of Melbourne, Victoria 3010, Australia
e-mail: zhiqfang@foxmail.com

© Springer Nature Singapore Pte Ltd. 2019
X. Qu et al. (eds.), *Smart Transportation Systems 2019*, Smart Innovation, Systems and Technologies 149, https://doi.org/10.1007/978-981-13-8683-1_6

lane and the adjustment factor of shared left-turn lane to estimate its SFR. Equation (6.1) is the calculation formula for the SFR of through lane and Eqs. (6.2) and (6.3) are the calculation formulas for the SFR of shared left-turn lane.

$$S_T = S_{bT} \times f_w \times f_g \times f_b \tag{6.1}$$

where S_T is the SFR for through lane; S_{bT} is a base SFR, usually 1900 passenger cars per hour per lane (pcu/h/ln); f_w is the adjustment factor for lane width; f_g is the adjustment factor for approach grade and heavy vehicles; and f_b is adjustment factor for the impact of bicycles.

$$S_{TL} = S_T \times f_{TL} \tag{6.2}$$

$$f_{TL} = \frac{1}{1.0 + 0.05 P_{LT}} \tag{6.3}$$

where S_{TL} is the SFR for shared left-turn lane; f_{TL} is adjustment factor for shared left-turn lane; and P_{TL} is proportion of left turns in lane group.

However, the SFR varies with the intersection geometry, channelization, weather condition, and ratio of heavy vehicles, which makes the adjustment method rather complicated. Therefore, HCM recommends the other method for SFR estimation, which is named field measurement method. This method needs to collect discharge headways at positions between the fourth vehicle and the last stopping vehicle in the queue when they pass through the stop line during green phase, and record the data for 15 signal cycles. The mean of these headways is used as saturation headways and then the SFR is calculated. However, there is a limitation in this method, since it assumes all discharge headways of queued vehicles are saturation headways. According to the field observation, discharge headways might be larger than saturation headways because of the distraction of drivers or slow start of some vehicles. Thus, the field measurement method cannot produce accurate estimation results for SFR.

Many researchers have devoted to the calculation of SFR. However, most of calculation methods are suitable for the single through lane, left-turn lane, and right-turn lane only. There are a few researches considering the SFR of mixed lanes. Shang et al. collected the headways of heterogeneous lanes during the middle and late time of the green phase and fit the distribution of each approach lane's headway. The sample headways in the 95% confidence interval are recommended to calculate the SFR of each lane [6]. Arasan and Vedagiri established the software package HETEROSIM and estimated saturation flow rates under heterogeneous condition. It has been found that there is a significant increase in the saturation flow rate with increase in the width of approach road, but the study is limited to estimation of saturation flow rate of straight-on traffic only [7]. Hamad and Abuhamda attempted to estimate the base SFR using the headways of 1,431 through-moving vehicles form 86 queues at three signalized intersections. They found that the result was clearly higher than the 1900 pcu/h/ln suggested by the HCM [8]. Stokes et al. adopted the data from 14 intersection approaches with exclusive double left-turn lanes to illustrate the use of simple linear regression to estimate saturation flows [9].

This paper uses KPSS test and HCM method to estimate the SFR of shared left-turn lane based on the field measurement headway data, respectively. KPSS test method divides the headways into four categories. They are the (i) headway between two going through vehicles, (ii) headway between going through vehicle and left-turning vehicle, (iii) headway between left-turning vehicle and going through vehicles, and (iv) headway between two left-turning vehicles. At first, the abnormal headways of each category because of the distraction of drivers and so on are eliminated until the remaining headways satisfy KPSS test. The estimated value of saturation headway is calculated as the weighted average of four categories' headways and then the SFR can be obtained. The results from two methods are compared in order to validate KPSS test is more accurate than traditional HCM method in the SFR calculation.

6.2 Data Collection

To validate the effectiveness of KPSS test method for SFR estimation for shared left-turn lane, a practical signalized intersection in Qu Jing, Yun Nan, China is taken as an example and the results are compared with traditional HCM method.

Qu Jing is the second largest city in Yun Nan Province with 130 km^2 metro area and 0.68 million of population. Qilin Eastern Road and Nanning Northern Road located in CBD are both urban arterial road. The intersection of the roads is controlled by signal lamps. Figure 6.1 is the layout and signal phase plan of the intersection. Each direction contains five entrance lanes, which are one right-turn lane, two through lanes, one shared left-turn lane, and one left-turn lane, and the number of exit lane is three. The average width of each lane is 3.5 m. The intersection has four signal phases during peak and off-peak hour, and the four entrances display green light in turn. In this paper, the shared left-turn lane of northern entrance has been selected as analytic target. The signal cycle of peak and off-peak hour is 212 s and 165 s, respectively. During peak hour, the green time of east, west, south, and west entrance are 40, 52, 48, and 60 s, while the green times are 26, 42, 35, and 50 s during off-peak hour. Having the surveillance video from traffic police department, the paper counted the headway of northern entrance shared left-turn lane manually by defining headway as the time difference between two successive vehicles' tails reaching the stop line. The time of data collection is 15:00–18:00, November 29, 2017. There are 811 valid data which has been collected with 408 at off-peak hour (15:00 ~ 16:30) and 403 at peak hour.

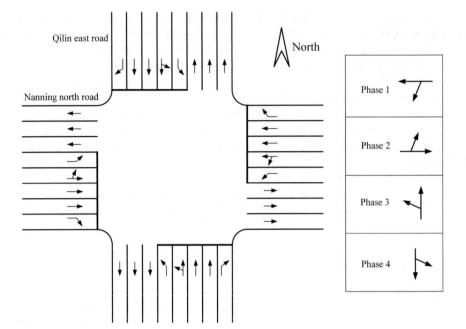

Fig. 6.1 The layout of the survey intersection and phase plan

6.3 SFR Estimation Result Analysis

Tables 6.1 and 6.2 show the headways during peak hour and off-peak hour based on KPSS stationarity test (SH stands for saturation headway).

Table 6.1 The calculation results based on KPSS test method during off-peak hours

Headway type	i	ii	iii	iv
Number of samples	67	55	78	175
Mean value of SH (s)	2.95	3.01	3.45	3.64
Standard deviation of SH (s)	1.07	1.21	1.99	2.88
Right side of confidence interval of SH (s)	2.69	2.68	3.00	3.21
Left side of confidence interval of SH (s)	3.20	3.34	3.90	4.06
Mean value of SFR (pcu/h)	1221	1194	1042	990
Right side of confidence interval of SFR (pcu/h)	1123	1077	923	885
Left side of confidence interval of SFR (pcu/h)	1340	1341	1198	1123
Total number of samples	375			
Weighted SH (s)	3.38			
Weighted SFR (pcu/h)	1064			

Table 6.2 The calculation results based on KPSS test method during peak hours

Headway type	i	ii	iii	iv
Number of samples	82	72	81	142
Mean value of SH (s)	2.96	3.12	3.17	3.16
Standard deviation of SH (s)	1.40	1.49	1.46	1.80
Right side of confidence interval of SH (s)	2.65	2.77	2.84	2.86
Left side of confidence interval of SH (s)	3.26	3.47	3.49	3.45
Mean value of SFR (pcu/h)	1218	1153	1136	1140
Right side of confidence interval of SFR (pcu/h)	1103	1037	1031	1041
Left side of confidence interval of SFR (pcu/h)	1340	1340	1198	1123
Total number of samples	377			
Weighted SH (s)	3.11			
Weighted SFR (pcu/h)	1158			

It can be seen from Table 6.1 that after abnormal data elimination, during off-peak hour, there are 375 samples remained and the saturation headways for four categories are 2.95, 3.01, 3.45, and 3.64 s. The weighted headway is 3.38 s and SFR is equal to 1064 pcu/h. While during peak hour, there are 377 samples that remained with 3.11 s weighted headway and SFR of 1158 pcu/h. According to Tables 6.1 and 6.2, it is noticed that the SFR difference between peak and off-peak hours at northern entrance shared left-turn lane is 94 pcu/h, which is 8.83%. That is because during peak hour, the intersection is more congested. Vehicles turn to have longer waiting time and drivers are concentrated on driving or rather impatient. Their vehicles will follow the front cars to form saturation flows. Whereas during off-peak hour, the intersection is not congested. Drivers are rather relaxed and during green time, the headways between front and behind vehicles become larger and the SFR becomes smaller.

Table 6.3 shows the estimated SFR of northern entrance shared left-turn lane by HCM method. During peak hour, the estimation value is 1144 pcu/h, a difference of 14 pcu/h, 1.2% compared with Table 6.2. Moreover, during off-peak hour, the estimation value is 1022 pcu/h, a difference of 42 pcu/h, 4.1%. According to Table 6.3, SFR during peak hour is higher than off-peak hour which is similar to the result considering Tables 6.1 and 6.2. Furthermore, during off-peak hour, there is a larger difference than during peak hour. That is because during off-peak hour, some of drivers are much more relaxed on driving and the vehicles form unsaturated flows. Those vehicles should be eliminated from estimating SFR while HCM method regards them as saturation headways which causes a lower SFR estimation value. However, during peak hour, drivers are more concentrated on driving. Usually, vehicles in queue could form saturation flows. At this time, the accuracy of HCM estimation increases and compared with the KPSS test method, the differences become smaller.

Table 6.3 Estimation results based on HCM method in the two investigation periods

Investigation period	Off-peak hour	Peak hour
Number of samples	408	403
SH (s)	3.52	3.14
SFR (pcu/h)	1022	1144

6.4 Impact of SFR Estimation on Signal Timing Plan

The value of SFR is directly related to the calculation of green time for each phase of signalized intersections. Furthermore, the traffic efficiency is affected as well. Section 6.3 analyzes and compares two methods for SFR estimation. During off-peak time, the difference is 1.2% while during peak hour, the difference is 4.1%. Although there is not much difference between two methods, it has significant influence on timing plan of signalized intersections. In this section, it is expressed by an actual intersection. Taking the intersection shown in Fig. 6.1 as an example, the paper calculates the green time of each phase, analyzes the impact of SFR error on intersection timing plan by comparing the SFR estimation result between the KPSS test method and HCM method.

According to the Highway Capacity Manual (HCM), the cycle length C of intersection can be calculated according to Eq. (6.4) as given below:

$$C = \frac{1.5L + 5}{1 - \sum_{i=1}^{4} (q_i / S_\mu^i)} \tag{6.4}$$

In the equation:

L (s)   total loss of green time of the intersection.
S_μ^i (pcu/h)   SFR of a key lane at i phase.
q_i (pcu/h)   arriving rate at i phase of a key lane.

The green time at phase i g_i is

$$g_i = (C - L) \times (q_i / S_\mu^i) / Y \tag{6.5}$$

All the key lanes of the intersection shown in Fig. 6.1 are shared left-turn lanes. L is equal to 12 s. Tables 6.4 and 6.5 provide the arriving rate and timing plan during peak and off-peak hours. It can be seen from Table 6.4, based on HCM method, the signal cycle length is calculated as 189 s, and green time is 38, 50, 44, and 57 s for each phase. Whereas based on the SFR estimation method of KPSS test, the signal cycle is 174 s with green time of 35, 46, 41, and 52 s for each phase. That is to say, although there is only 14 pcu/h difference between the SFR estimation using the two methods, the difference in cycle length is 15 s, which is 7.9%. Similarly, as shown in Table 6.5, the difference in SFR is 42 pcu/h but the difference in cycle

Table 6.4 Timing plans of the intersection under two SFR estimation methods in peak hour

Phases	q_i (pcu/h)	HCM method			KPSS test method		
		S_μ^i (pcu/h)	g_i (s)	C (s)	S_μ^i (pcu/h)	g_i (s)	C (s)
1	200	1144	35	189	1158	32	174
2	264		47			43	
3	238		41			38	
4	303		54			49	

Table 6.5 Timing plans of the intersection under two SFR estimation methods in off-peak hour

Phases	q_i (pcu/h)	HCM method			KPSS test method		
		S_μ^i (pcu/h)	g_i (s)	C (s)	S_μ^i (pcu/h)	g_i (s)	C (s)
1	150	1022	28	161	1064	22	130
2	240		44			36	
3	200		37			30	
4	286		52			42	

length is 31 s, the difference ratio reaches 19.2%. In this case, it is believed that a small difference in SFR estimation could cause a dramatic change in timing plan of signalized intersections. So, a more accurate SFR estimation value would be significant.

Signal timing plan has an important influence on the average delay of intersection. Particularly the delay will increase rapidly due to a longer signal cycle. According to flow information, signal timing plan is shown in Tables 6.4 and 6.5, and the average vehicle delay at the intersection could be calculated. During peak hour, the average delay is 89.8 s using HCM method and 82.8 s using KPSS test method. The delay increases by 7.8% because of the 4.1% difference in SFR estimation value. During off-peak hour, the delay is 74.2 s and 63.8 s, respectively, which means due to the 1.2% difference in SFR estimation value, the average vehicle delay at intersection increases by 14.0%.

6.5 Conclusion

The paper compares KPSS test method with HCM method in the SFR estimation for shared left-turn lane and then analyzes the influence on signal timing plan and vehicle delay by SFR estimation error. The results show that

I. Because of distracting driving during green time, those vehicles have larger headways when following their front vehicles and form an unsaturation flow. Those data should be eliminated from SFR estimation. Especially, drivers prefer to use mobile phone during red time even until the signal lamps turn green and the front car is starting. The headways increase under this circumstance. It is a common phenomenon at signalized intersections in modern life.

II. The accuracy of SFR estimation has significant influence on the signal timing plan and average vehicle delay. During peak hour, using HCM method, the estimated SFR is 4.1% lower than using KPSS test method. However, that causes a 7.9% increase in cycle length and 7.8% increase average vehicle delay. During off-peak hour, the SFR estimation value is 1.2% lower and that causes a 19.2% increase in cycle length and 14.0% increase in average vehicle delay.

The SFR at the signalized intersection is affected by other factors, such as weather condition, number of lanes, and the ratio of buses [10–13]. In the next step, the authors will study these contributing factors based on field data.

Acknowledgements This study is supported by the National Natural Science Foundation of China (No. 71771062 & 71771050).

References

1. Qu, X., Zhang, J., Wang, S.: On the Stochastic fundamental diagram for freeway traffic: model development, analytical properties, validation, and extensive applications. Transp. Res. Part B Methodol. **104**, 256–271 (2017)
2. Zhou, M., Qu, X., Li, X.: A recurrent neural network based microscopic car following model to predict traffic oscillation. Transp. Res. Part C Emerg. Technol. **84**, 245–264 (2017)
3. Li, X., Medal, H., Qu, X.: Connected heterogeneous infrastructure location design under additive service utilities. Transp. Res. Part B Emerg. Technol. **120**, 99–124 (2019)
4. Zhang, J., Qu, X., Wang, S.: Reproducible generation of experimental data sample for calibrating traffic flow fundamental diagram. Transp. Res. Part A Policy Pract. **111**, 41–52 (2018)
5. Highway Capacity Manual.: Transportation Research Board. Washington, DC (2000)
6. Thamizh, A.V., Vedagiri, P.: Estimation of saturation flow of heterogeneous traffic using computer simulation. In: Proceedings 20th European Conference on Modelling and Simulation (2006)
7. Shang, H., Zhang, Y., Fan, L.: Heterogeneous lanes' saturation flow rates at signalized intersections. Procedia Soc. Behav. Sci. **138**, 3–10 (2014)
8. Hamad, K., Abuhamda, H.: Estimating base saturation flow rate for selected signalized intersections in Doha, Qatar. J. Traffic Logist. Eng. **3**(2), 168–171 (2015)
9. Stokes, R.W., Stover, V.G., Messer, C.J.: Use and effectiveness of simple linear regression to estimate saturation flows at signalized intersections. Transp. Res. Rec. **1091**, 95–101 (1986)
10. Bie, Y., Liu, Z., Ma, D., Wang, D.: Calibration of platoon dispersion parameter considering the impact of the number of lanes. J. Transp. Eng. **139**(2), 200–207 (2013)
11. Wang, Y., Bie, Y., An, Q.: Impacts of winter weather on bus travel time in cold regions: case study of Harbin, China. J. Transp. Eng., Part A Syst. **144**(11), Article number: 05018001 (2018)
12. Wang, Y., Bie, Y., Zhang, L.: Joint optimization for the locations of time control points and corresponding slack times for a bus route. KSCE J. Civ. Eng. **23**(1), 411–419 (2019)
13. Bie, Y., Gong, X., Liu, Z.: Time of day intervals partition for bus schedule using GPS data. Transp. Res. Part C Emerg. Technol. **60**, 443–456 (2015)

Chapter 7
Safety on the Italian Highways: Impacts of the Highway Chauffeur System

Serio Agriesti, Luca Studer, Paolo Gandini, Giovanna Marchionni, Marco Ponti and Filippo Visintainer

Abstract Road safety is certainly a main drive toward automated driving. Removing the human factor from driving maneuvers allows to prevent crashes caused or fostered by human flaws, such as distraction or tiredness. However, full automation on public roads is still a goal to be achieved, and different solutions are currently being developed. In the short term, one of the most promising examples is the Highway Chauffeur system, capable of performing fully automated driving in defined conditions and environments. The scope of this paper is to evaluate the number of crashes that could possibly be addressed through a defined market penetration of Highway Chauffeur vehicles (HC vehicles in the article). The evaluation is based on a bibliographical review and on an analysis of crash-related data recorded in Italy by the National Statistical Institute—ISTAT—on public roads. By using the information publicly available on ISTAT's website, this paper provides a first magnitude concerning the number of crashes addressable by the Highway Chauffeur system, and final considerations on the future research to be carried out.

7.1 Introduction

The Highway Chauffeur (HC) system allows drivers to exit the driving loop and entrust the driving task to the vehicle, until some conditions are met. These conditions define the Operational Design Domain (ODD) of the system, namely, where and when the automated system is able to guarantee a safe and efficient driving. The elements generally defining an ODD are provided in [1], such as roadway typologies, geographical areas, speed ranges, environmental conditions, etc. A Highway Chauffeur system is intended to operate on highways, in a speed range of $0 \div 130$ km/h [2], as long as the equipped perceiving system [3] is not hindered by adverse weather

S. Agriesti (✉) · L. Studer · P. Gandini · G. Marchionni · M. Ponti
Mobility and Transport Laboratory, Politecnico di Milano – Dipartimento di Design, Milan, Italy
e-mail: serioangelo.agriesti@polimi.it

F. Visintainer
Centro Ricerche FIAT, FCA EMEA Product Development, Safety and Driver Assistance Systems, Infotelematic Systems, Trento, Italy

© The Author(s) 2019
X. Qu et al. (eds.), *Smart Transportation Systems 2019*, Smart Innovation, Systems and Technologies 149, https://doi.org/10.1007/978-981-13-8683-1_7

conditions such as heavy rain, snow, or fog. Moreover, in order to operate, the system requires well-defined lane markings and readable vertical signals; besides, it can benefit from information on the state of the road surface (e.g., wet surface, black ice). This information can be drawn both by dedicated sensors and through V2X communications [4]. As long as the system is engaged, the driving of an HC vehicle can be considered automated, with the driver completely out of the driving loop. Therefore, it is safer and more efficient than the driving of a human driver. Indeed, a sustainable deployment of the autonomous driving technology in future society can happen only if the technology itself reaches at least the same level of safety and efficiency currently achieved by a traditional vehicle. This aspect is essential for a safe and successful phase in of automated driving.

Taking into consideration the performance of the Highway Chauffeur system reported in literature, this paper explored the potential impact on safety of HC vehicles in the traffic flow of the Italian highways. This task was accomplished by analyzing crash data recorded by ISTAT in 2016, available at www4.istat.it/it/archivio/87539 [5]. The goal of the analysis presented in the paper is to define which crashes occurred in scenarios included in the most likely ODD of an HC vehicle, and which were caused or fostered by elements referable to automation. The approach of this study was based on the methodology presented in [6], even if with some expected differences, resulting from the adoption of two different databases for input data. In fact, [6] is a study conducted on the target crash population in the USA, while this paper considered the Italian framework on the basis of the data recorded by ISTAT. As a result of such analysis, this research provides an estimation of the percentages of potentially avoidable crashes for a 10% level of market penetration of the Highway Chauffeur with regard to the Italian highway network (according to [7], providing a forecast of the L3 market penetration in year 2025, 10% was considered a plausible value). This result aims to provide an overall estimation of the magnitude of the future impacts of the Highway Chauffeur system on the Italian highways. Indeed, the data analysis can be considered as an aggregate study and does not provide all the elements capable of retracing and characterizing every single crash event. Therefore, when the accident dynamics were uncertain, the accident was considered unavoidable and not addressable. This approach is conservative, and it probably underestimates the real impacts of the Highway Chauffeur system. Nonetheless, it meets the objective of providing a minimum baseline for potential safety impacts that will most likely be obtained in the medium term. In fact, the purpose of this paper is to identify crashes that can be addressed through the use of the Highway Chauffeur system, obtaining an overall value that represents both crashes potentially avoidable and crashes whose severity can be reduced. This kind of analysis is characteristic of an Ex-Ante phase, during which the new transport system is not yet implemented on roads but a first evaluation is needed to assess the magnitude of potential results and to have elements to design evaluation activities and field tests.

The paper is structured as follows: Section 7.2 provides an overview of the Highway Chauffeur's functioning and capabilities, defining the typical ODD. Section 7.3 presents the dataset employed in the analysis; it also analyses the most relevant fostering elements and causes connected to the recorded crashes when the dynamics resulted reasonably defined to evaluate the probable impact of automated HC vehicles (in analogous situations); lastly, it provides figures aimed at obtaining an impact assessment of the Highway Chauffeur system on safety. Section 7.4 provides comments on the results obtained and additional considerations on the assessment carried out. Finally, Sect. 7.5 presents the conclusions of the study and indicates future research directions, especially those considered useful to improve the estimation provided and make the magnitude more quantitative rather than indicative.

7.2 Overview of the Highway Chauffeur System

HC vehicles can perform both overtaking and lane changing [8–10] with different levels of aggressiveness [11, 12]. Moreover, when facing congestion, the Traffic Jam Chauffeur function guarantees a safe and efficient driving during the Stop&Go regime. To allow the driver to engage secondary tasks, the Highway Chauffeur system exploits onboard equipment that includes Adaptive Cruise Control (ACC), Lane Keeping Assistant, and Automated Braking System, just to mention a few. An essential role is thus played by the software and hardware components [3]. With regard to the longitudinal control, while in automated mode, the HC vehicle relies on the ACC to keep a cruise speed value or a safe distance/time gap from the vehicle in front of it. To accomplish this task, long- and short-range radars onboard allow the system to compute the safe and desired time gap on the basis of the speed difference and the distance between the HC vehicle and the one in front of it. At the same time, the perceiving system reacts faster than drivers to the braking of the vehicle in front of it, bypassing almost completely the human perception and reaction time [13]. Moreover, giving the automated system a longitudinal control ensures that the acceleration and deceleration regimes do not suffer from the driver's tiredness or distraction (typical of human drivers), thus reducing the number of harsh braking and the possible resulting collisions.

As mentioned, the HC system can also accomplish lane keeping, lane change, and overtaking without the need to re-engage the human driver. To manage this task, the perceiving system employs the onboard sensors to measure both the available spaces and the driving speed on the target lane, deriving safe time gaps to perform the maneuver. In this paper, the assumption is that the HC vehicle is capable of identifying the available gaps on the other lane and to act accordingly.

7.2.1 Approach

Based on the literature about the HC system, it is possible to analyze crash-related data to determine what element caused or fostered crashes and if these occurred in scenarios ascribable to the typical ODD. Therefore, if the triggering or fostering factor is addressable by automated driving, a first evaluation of the number of avoidable or mitigated crashes is provided. These crashes can be defined as Target crashes (as in [14]) representing "the maximum potential safety benefit if the fully deployed system was 100% effective in reducing target crashes." An example referring to the HC system is a crash caused by the driver's distraction, occurred on a highway, in a section with clear lane marking and signaling, dry pavement and fair weather. An hypothesis imposed by the features of the data available concerned weather conditions: ISTAT's public database does not consider different levels of severity for adverse weather conditions such as, for example, rain. Thus, the only choice was to consider each of these events out of the ODD system (that, actually, could endure light rain). The same consideration was made for snow, fog, and hail. Thus, the accidental events occurred clearly within the ODD's boundaries were analyzed, in order to give a more realistic estimation. A similar approach was followed in [14], investigating V2I communications: "Targeted crashes are crashes that could potentially be eliminated through the deployment of a specific V2I application or set of applications (i.e., researchers should determine the potential benefit of an application area, assuming 100% effectiveness and 100% deployment)." It should also be highlighted how the analyses carried out in [14] are based on the identification of pre-crash scenarios, similarl to what was accomplished in this paper. Another research, within the activities of the U.S. Department of Transportation and adopting a similar approach, is provided in [15], designed with the objective to estimate "the upper limit of annual police-reported crashes that could potentially be addressed with IntelliDrive safety systems based on vehicle-to-vehicle communications or vehicle-to-infrastructure cooperation." Even in this work, the result is a magnitude rather than a figure, as in this paper. This choice seems to be forced by the limited number of field operational tests on public roads characterizing both cooperative and autonomous vehicles. Moreover, it can be often difficult to assess the actual impacts on safety also through the Field Operational Test, due to the limited number of actual crashes. Therefore, this paper should be considered as a first tool for future evaluations carried out by road authorities, safety organizations, or researchers and evaluators. Indeed, it provides a first step useful to highlight which crashes are most likely preventable or reducible in severity thanks to an automated or semi-automated vehicle.

In the framework of road safety theories, the presented approach finds itself within the "Causal accident theory" category and as an accident causation approach tries mostly to consider the direct in-vehicle modification of the driving task [16]. The direct in-vehicle modification is indeed the main safety mechanism triggered by the HC system. Of the other safety mechanisms listed in [16], the following analysis considers also the modification of accident consequences while some relevant subjects are highlighted as future research topics but not included in the analysis (mostly due to the limits of the literature and of crash-related data on Italian roads). Of these safety mechanisms, the ones that will be mostly influenced by a certain market penetration of L3 vehicles are as follows:

- The indirect modification of user behavior. This mechanism concerns mostly the Take-Over maneuver during which the human driver takes the control of the vehicle back. That maneuver falls outside the set of crashes analyzed in Sect. 7.3 and thus doesn't affect the overall analysis.

- The long-term behavioral adaptation of the human driver. This mechanism should impact mostly crashes outside the considered ODD. Still, preferences about the chosen time gap or the acceleration regime of the L3 vehicle (when engaged) are questions that don't find a univocal answer in the literature.
- Modification of road user exposure and of modal choice. These two mechanisms are the ones that will probably affect the most the presented analysis. It is clear that a certain level of crash reduction can be outclassed by a relevant increase of kilometers driven indeed (due to the reduced cost of travel time, while the vehicle drives itself and the human driver is allowed to do something else). The analysis reported in this paper shouldn't be relevantly affected by an increase of road user exposure due to the short time horizon and the low market penetration considered. In fact, the changes of the travel patterns will probably arise in the medium-long term and currently impossible to quantify.

7.2.2   Equipment Overview and Effectiveness of the System

In order to bound the capabilities of the hypothesized HC system to the analysis carried out in paragraph 3, a short overview of the equipment of an L3 vehicle is provided. Moreover, as in [17], each one of the following components is related to the kind of crash that it addresses. Also, it should be noted that some of the crashes are not addressable by one single component but can be addressed by the system as a whole (through data fusion and software processing).

An L3 vehicle on the market by the year 2025 is considered equipped at least of Improved Advanced Cruise Control [18], Lane Keeping Assist, Automated Braking System, Front Collision Warning, Lane Change Assistance, and Blind Spot Detection System. All these components can be related to addressable crash types, as, for example, in [17]. For the present analysis, the following clusters are considered:

- Improved Advanced Cruise Control + Forward Collision Warning + Automated Braking System: head crash, lateral-head crash (within the cut-in scenario), rear end, and collision against an obstacle.
- Lane Keeping Assist + Lane Change Assistance + Blind Spot Detection System: lateral crash, road departure, and collision against an obstacle.

As mentioned, these single components can usually address the corresponding crashes with a level of effectiveness or eventually through the driver intervention (that is also characterized by a certain effectiveness). For an L3 vehicle, additional elements should be considered though, namely, onboard sensors, data fusion, and software processing. Through these elements, the hypothesis of 100% effectiveness can be adopted without compromising the results of the analysis. In fact, the first relevant difference between an automated system and a human driver is the perception capability. The automated system drives the vehicle mainly thanks to the sensors suite and the software. The minimum equipment in this case is composed of long- and short-range radars, LIDAR, and cameras. Each of these components has strengths and weaknesses, mostly covered by the other onboard sensors as explained in [19]. The hypothesis in this case is that, by 2025, the sensor suite is going to correctly perceive the surroundings of the vehicle with no system failures, as long as the ODD conditions are met, reaching a full awareness. This is the first, relevant, difference between an L3 system and a human driver and is expected to influence all the crashes (analyzed in Sect. 7.3) in which an incorrect perception of the surroundings presumably took an important role in the crash. Through the perception process, once each sensor recorded information, the software is in charge to process all the inputs and to define a trajectory. This is the second, relevant, advantage of the HC on a human driver because the software does not suffer from tiredness or distraction, it is not indecisive and it can keep up the same level of performance even after hours of driving. As it will be deepened, these factors contribute to a really high percentage of crashes. Moreover, [17] is a work that summarizes the effectiveness of Driving Assistance systems and Connected Vehicle technologies on the basis of a bibliographical review.

Another work that achieves a similar task is [20], from which another valuable insight on the relationship between pre-crash scenarios and driving assistance system was obtained. Moreover, also in this work, a set of coefficients expressing the effectiveness of these technologies in addressing crashes was provided.

Unfortunately, after careful consideration it was decided to not use the coefficients arising from [17, 20] in this paper, mainly because all the coefficients refer to L0 or to L1 systems and quantify the ability of a single component to address a certain pre-crash scenario (this reflects also the current state of the art and the available bibliography). This means that the contributions of the sensors' suite and of the onboard software are ignored, even though they are the ones that nullify fostering factors such as distraction or tiredness in the pre-crash scenarios. Thus, it would be extremely conservative and also incorrect to apply these coefficients to a 2025 L3 vehicle, even by combining the contributions of all the technological components onboard. This imposes the hypothesis of 100% effectiveness, which is a limitation difficult to overcome only with the current literature. Moreover, the amount of field tests currently carried out in Europe is still limited and won't provide enough data to improve this type of analysis on a crash database. Thus, the need for surrogate methodologies arises and this kind of safety assessment could strongly benefit from analyses such as the one carried out in Sect. 7.3.

7.3   ISTAT's Database—2016 Microdata

ISTAT is the Italian National Statistical Institute that provides, among others, public data on road accidents involving injured or dead people. The macro-data for public use concerning the entire national network was considered for a first evaluation of crash scenarios potentially addressable by the Highway Chauffeur system when the driving task is entrusted to automation. Therefore, an analysis was conducted on the crashes that occurred in 2016 on the Italian public roads. The dataset used includes the following details (only the relevant ones are listed below):

"Autonomous vehicle based systems are also considered to evaluate what additional safety enhancements they can effect. (…). Applicable crashes include rear-end crashes, lane departures, lane change or merge crashes, curve speed or excessive speeding crashes, and stop sign violations. It is assumed that AV systems could potentially address pedestrian, cyclist, and animal crashes as well as loss of control, road departure, and maneuver crashes in which speeding is a contributing factor" [15].

A first selection was carried out among all the crashes that occurred considering only the crashes occurred on highways (amounting to 9,360 in 2016), since the HC function can only prevent accidents on the type of road for which it is designed. The research then considered only the events (amounting to 6,408 crashes in 2016, 69% of the 9,360 occurred) that presumably occurred in domains included in the typical ODD (as derived from literature). On the basis of the data available, this first selection considered the features listed in Fig. 7.1: paved road surfaces, dry roads, both vertical and horizontal signals, and clear weather conditions. Considering the resulting set of crashes, evaluations were made concerning the circumstances that fostered or caused the events in order to obtain the number of accidents that could have been prevented or mitigated in severity if one of the two vehicles had been driven automatically (Fig. 7.2). It should be mentioned that the HC vehicle was considered capable of passing through roadworks [21], thus the temporary signaling was not considered sufficient to discard this dataset (equal to around 1%). The details on the circumstances considered as cause or contribution to the crash occurrence are presented below.

Distracted driving or indecisive behavior. When the HC system is engaged, it continuously scans the environment and the surrounding road users, constantly uploading the relevant information without getting tired or distracted. Moreover, as long as the HC vehicle finds itself in its ODD, no undecided behavior should arise. In fact, an automated system decides in a short timeframe, almost instantaneously. Exceptions to this assumption can concern system failures, which are not considered in this paper (being extremely bound to the control algorithms of each OEMs and also rare enough not to impact the evaluation). Considering all the crashes that occurred in the specified ODD, 1,291 were caused or fostered by distracted driving or an indecisive behavior. Potentially, they all appear possibly avoidable thanks to the HC system. From an in-depth analysis, it appears that

- 839 crashes occurred with no additional cause or fostering element ascribable to the other vehicle involved. Thus, they are all addressable by an HC vehicle and can also be considered avoidable, considering that the human flaw would be neutralized. 623 of these events were rear-end collisions which an HC system should be able to address and very likely prevent.
- 199 collisions occurred while both drivers involved were distracted or showed an indecisive behavior. 144 of these were rear-end collisions, thus probably avoidable when the HC vehicle is the one behind. With the data available, it cannot be reasonably determined what would happen if the HC vehicle were the one in

Tested CV&DA technologies and corresponding pre-crash scenarios.

CV&DA Technology	Automation Level(SAE)	Target Pre-Crash Type and Pre-Cash Scenarios
NS: Forward Collision Warning (FCW)	0	Rear-End:
NS: Autonomous Emergency Braking (AEB)	1	1 Lead Vehicle Stopped
NS: Collision Warning System (CWS)	0	2 Following Vehicle Making a Maneuver
S: Forward Collision Warning (FCW) + Adaptive Cruise Control(ACC)	1	3 Lead Vehicle Decelerating
S: Forward Collision Warning (FCW) + Autobrake	1	4 Lead Vehicle Moving at Lower Constant Speed
S: Forward Collision Warning (FCW) + Autonomous Emergency Braking (AEB)	1	5 Lead Vehicle Accelerating
S: Adaptive Cruise Control(ACC) + Advanced Braking System (AdvBS)	1	
S: Adaptive Cruise Control(ACC) + Advanced Braking System (AdvBS) + Collision Warning System (CWS)	1	
S: Collision Mitigation Brake System (CMBS)	1	
NS: Pedestrian Crash Avoidance and Mitigation System(PCAM)	1	Pedestrian:
		1 Pedestrian Crash With Prior Vehicle Maneuver
		2 Pedestrian Crash Without Prior Vehicle Maneuver
NS: Blind Spot Warning (BSW)	0	Lane Change:
NS: Lane Change Warning (LCW)	0	1 Vehicle(s) Turning – Same Direction
S: Blind Spot Warning (BSW) + Lane Change Warning (LCW)	0	2 Vehicle(s) Changing Lanes – Same Direction
		3 Vehicle(s) Drifting – Same Direction
NS: Intersection Movement Assist (IMA)	0	Crossing Paths:
		• Vehicle Turning Right at Signalized Junctions
		• Vehicle Turning at Non-Signalized Junctions
		• Straight Crossing Paths at Non-Signalized Junctions
		• Running Stop Sign
		• Running Red Light
NS-CV: Left Turn Assist(LTA)	0	Crossing Paths:
S: Collision Mitigation Brake System (CMBS)	1	• Left Turn Across Path from Opposite Directions at Non-Signalized Junctions
		• Left Turn Across Path from Opposite Directions at Signalized Junctions
NS: Lane Departure Warning(LDW)	0	Run-Off-Road:
S: Lane Departure Warning(LDW) + Curve Speed Warning(CSW)	0	• Road Edge Departure With Prior Vehicle Maneuver
		• Road Edge Departure Without Prior Vehicle Maneuver
		• Road Edge Departure While Backing Up
NS: Electronic Stability Control (ESC)	1	Run-Off-Road:
		• Control Loss without Prior Vehicle Action
		• Control Loss with Prior Vehicle Action
NS: Backup Collision Intervention (BCI)	1	Backing:
NS: Rearview Cameras (RCA)	0	• Backing Up into Another Vehicle

*Note:*S stands for the integrated system, while NS stands for non-syste.

Fig. 7.1 Tested CV&DA technologies and corresponding pre-crash scenarios [17]

Road Type	According to the Italian classification
Road Section	• **Crossing** • **Intersection** • **Curve** • **Roundabout** • **Straight Road**
Road Surface	• **Paved** • Paved – Dismaled • Unpaved
Surface State	• **Dry** • Slippery • Snow • Wet • Frozen
Signage	• Absent • **Orizzontal** • **Both vertical and horizontal** • **Vertical** • **Roadworks**
Weather Condition	• **Clear** • Rain • Snow • Fog • Hail
Kind of crash	• **Rear-end collision** • **Lateral collision** • **Etc.** • **Head-on collision** • **Skid**
Kind of vehicle involved	• **Car** **vehicle** • **Bus** • **Emergency** • **Truck** • **Motorcycle**
Circumstance that fostered the event	Reported in the analysis

Fig. 7.2 ISTAT's relevant data—in bold the data considered for the analysis

front, though it can be reasonably asserted that an HC vehicle performs a shorter and smoother braking. Nevertheless, it cannot be assessed in how many of these accidents the driver of the vehicle behind was so distracted or indecisive that the rear-ended crash would have occurred regardless of the harshness of the braking of the vehicle in front. Therefore, these 144 events were not considered among the ones addressable by an HC vehicle. 13 collisions were lateral-frontal crashes potentially avoidable if the HC vehicle performed the lane-changing maneuver, adequately scanning the other lane as mentioned in Sect. 7.2. They can be considered addressable also when the HC vehicle is the one behind: even if the preceding vehicle performs a risky lane change, the HC vehicle should be alerted and ready to decelerate. This consideration is based on the assumption that the ACC of the HC vehicle receives more information through the enhanced perceiving system of an L3 vehicle and thus reacts faster and better than a standard ACC [18].

- 22 collisions occurred because the other vehicle involved did not keep a safe distance from the one in front of it. This kind of crash is most likely addressable because the HC vehicle keeps a safe distance from the one in front and avoids harsh braking, thus lowering the chances of the vehicle behind to rear end.
- In seven cases, the other vehicles involved were driving at a speed higher than the speed limit. In six of these cases, the accident resulted in a rear-end collision. It should be noted that the HC vehicle brakes in a smoother way compared to a possibly distracted human driver, making a rear end less likely. Since the dynamics of the crash are not clear, these cases were not considered among the ones addressable.
- 35 collisions took place against an obstacle on the road, five of which during a turn. It should be noted that the HC system should be able to detect this obstacle while in its ODD and prevent the collision. Such collisions are thus addressable by the HC system.

In conclusion, out of the 1,291 collisions caused or fostered by distraction or an indecisive behavior, 1,136 (around 88%) resulted presumably addressable by the L3 system if at least one of the two vehicles involved is an HC vehicle. The number of crashes addressable in such case is equal to around 18% of the 6,408 included in the ODD. It should also be highlighted that, in the bullet list presented above, only some examples were reported and commented, while the same analysis was carried out for each one of the 1,291 scenarios recorded in ISTAT's database (Fig. 7.3).

Insufficient safety distance. As mentioned in Sect. 7.2, an HC vehicle driving with the system engaged relies on the ACC for a longitudinal control. This means that a safe time gap is always kept between the HC vehicle and the one in front of it. Therefore, the contribution of the safety distance to the crash event is removed from the equation. Considering only the ODD, 1,679 accidents were caused by an insufficient safety distance. From an in-depth analysis, it appears that

- 1,570 of the crashes occurred were not fostered by flaws ascribable to the other vehicles involved. Therefore, they are all considered addressable by the HC system and, thus, likely preventable. 1,285 of these events where rear-end collisions, which a Highway Chauffeur system should be able to address and very likely prevent.

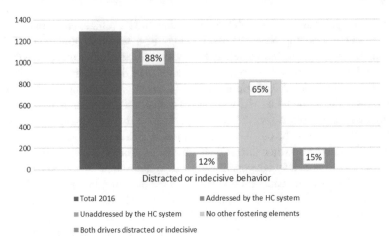

Fig. 7.3 Distracted driving or indecisive behavior

- 39 collisions occurred because the other driver was distracted or showed an indecisive behavior. Nevertheless, erasing the limited safety distance should make these crashes potentially avoidable or mitigated in their severity (especially because 31 of them were rear-end collisions).
- 36 collisions took place with the other vehicles because not keeping the safety distance. From the data available, it was not possible to determine the dynamics of these crashes, especially because it was not possible to assess how much the other vehicle involved concurred in the crash (i.e., insufficient safety distance). Therefore, adopting a conservative approach, all these crashes were judged as not addressable (or, to be more precise, they were not summed to the addressable ones).

Therefore, out of the 1,679 collisions caused or fostered by an insufficient safety distance, 1,643 (around 98%) can be addressed by an L3 system such as the Highway Chauffeur, as long as the vehicle involved is an HC vehicle. The number of crashes addressable in this case is equal to around 26% of the 6,408 included in the ODD, (since the crashes considered in the analysis were those included in the ODD, no factor such as a wet road could have concurred to the event). Again, it should also be highlighted that, in the bullet list presented above, only some examples were reported and commented, while the same analysis was carried out for each one of the 1,291 scenarios recorded in ISTAT's database.

Speeding. If the Highway Chauffeur system is engaged, the driving speed cannot exceed the mandatory limit. Therefore, many of the 422 accidents occurred in scenarios falling within the ODD described above can be surely considered addressable. From an in-depth analysis, it appears that

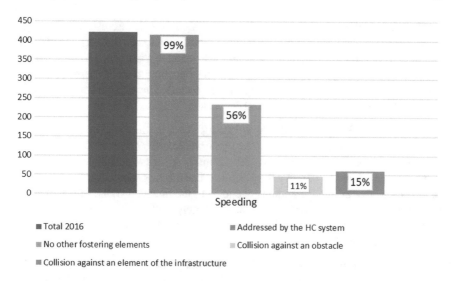

Fig. 7.4 Speeding

- 234 crashes occurred with no other fostering element apart from the speeding one, thus they are considered both addressable and most likely avoidable with an HC system.
- 24 collisions took place while the driver of the other vehicle appeared distracted or showed an indecisive behavior. This type of accident was considered addressable, as well as the 16 that were rear-end collisions, since the speeding contribution can be prevented.
- 60 collisions occurred against an obstacle on the road. These accidents were considered addressable by the L3 system and also potentially preventable.
- 21 collisions occurred against a regularly stopped vehicle. These accidents were considered addressable by the L3 system and also potentially preventable.
- 61 collisions occurred against an element part of an infrastructure. Even these accidents were considered addressable by the L3 system and also potentially preventable when erasing the speeding factor. It should be noted that the crashes considered in the analysis are the ones included in the ODD, thus no factor such as a wet road can concur to the event.

Out of the 422 collisions caused or fostered by distraction or an indecisive behavior, 416 (around 99%) resulted addressable and also presumably avoidable if one of the two vehicles involved is an HC. The number of crashes addressed in this case amounted to around 6.5% out of the 6,408 included in the ODD (Fig. 7.4).

Skidding or road departure due to distraction. As mentioned, having the system engaged should prevent crashes and events caused by a distracted behavior. 210 events occurred in the attempt to avoid an obstacle or a vehicle, presumably perceived by the human driver at the last moment. 75 events occurred without a recorded object to be avoided and, thus, were likely due to distraction alone. All these crashes (341 events) were considered addressable and likely preventable when the Highway Chauffeur is engaged. It is useful to remember that all these accidents happened in the intended ODD, thus with a dry road surface and no adhesion compromised (namely, with no other concurring factor in the dynamics of the event). This category represents 5% of the 6,408 events included in the ODD.

Skidding or road departure due to speeding. Also, in this case, the main external factors are the avoidance of an obstacle or a vehicle (297 events) and events not involving any object to be avoided (306 events). All these accidents (623 events) were considered addressable by the Highway Chauffeur system. Again, it is useful to remember that all these accidents occurred in the intended ODD, thus with a dry road surface and no adhesion compromised (namely, with no other concurring factor in the dynamics of the event). This category represents 10% of the 6,408 events included in the ODD.

Skidding or road departure to avoid an obstacle. These 155 events were not triggered by a vehicle's inappropriate behavior but still, while in its ODD, the HC vehicle should exploit its enhanced perception to reduce this type of event. Nevertheless, the dynamics of these crashes are not well defined by the data available and, therefore, were not considered addressable by the Highway Chauffeur system alone.

7.4 Analysis of the Results

The cases considered in Sect. 7.3 represent the more common ones, as far as causes and fostering elements are concerned, accounting alone for 70% of all the crashes occurred in scenarios ascribable to the ODD. It should also be considered that some crashes could not be analyzed in depth, such as the 431 cases in which no cause or fostering element is recorded for the vehicle and, thus, the dynamics of the crashes could not be recovered from the data available. Therefore, the share represented by the cases reported in the section above increases up to 76% of the analyzable events.

Around 4% of the crashes account for other scenarios, limited in numbers, that are not reported in this paper, not being statistically relevant (e.g., eight crashes in which the vehicle was overtaking without complying with the signaled overtaking ban were considered in the overall statistics, but not analyzed in depth in this paper). From the analysis of these cases, only 152 events ($\approx$0.02%) resulted as not addressable even with the automated system engaged. To obtain the most precise results possible, one last consideration should be made: 1,222 events occurred with the vehicle driving regularly, which means that the cause or the fostering element was probably due to the other vehicle involved. These events could have been avoided only with both vehicles equipped with the Highway Chauffeur system. Considering a market penetration of 10%, the odds of avoidance are equal to $\approx$1% (0.009) which means that around 12 crashes of this category result addressable thanks to the use of HC vehicles on the Italian highways. The percentage of 0.9% should actually be applied to every type of crash in which the causes are ascribable to the other vehicle involved. However, in this paper, it was applied only when the numbers resulted statically relevant (the chosen threshold was set at 1,000 crashes).

The value of 10% for the market penetration parameter was chosen for the safety assessment in a short-medium term in order to retain a certain congruence with the 2016 accident-related data. Moreover, said parameter was chosen also because, in the long term, both the likely reached level and the capability of automated driving are hardly comparable to the Highway Chauffeur system. Besides, in the upcoming years, the impact on the safety of other technologies, such as C-ITS services and V2X communications, should be analyzed in their own capability of preventing accidents, but also in the possibilities granted by the joint implementation of L3 vehicles and cooperative technologies (e.g., the C-ITS Use Case Adverse Weather Conditions that can expand the ODD of a Highway Chauffeur vehicle or, at least, make the control transition from the automated system to the driver safer). This Take-Over transition can prove to be the cause or the fostering element for new types of crashes. However, it was not considered in this analysis, since it is not included in ISTAT's database for obvious reasons. Moreover, new types of crashes could and probably will arise from the complex relationship between the two authorities onboard: the human driver and the Highway Chauffeur system. This field of evaluation was not included in the results presented in this paper. With regard to this topic, an increasing number of works can be found in literature. However, numerical results referring to national realities are still missing and probably will continue to be missing until the system is used on public roads or a sufficient number of field tests are carried out.

7.5   Conclusions and Future Works

Out of the total number of crashes occurred during 2016, around 70% could have involved an HC vehicle driving in automated mode (meaning that 70% of the crashes occurred in the ODD conditions). This percentage could grow if the capability of the system to face light adverse weather conditions is taken into account. It is important to highlight how, considering the single event involving a single HC vehicle, the percentages of presumably avoidable crashes are rather high: 78% involving a distracted or an indecisive behavior, 98% involving an insufficient safety distance, and 99% when speeding is considered. It should be highlighted that the main impact is due to the incapability of the system to break traffic rules, thus abiding by safety distances and speed limits. However, it is also worth noticing how, in 2016, speeding and unsuitable safety distances accounted for around 33% of the crashes occurred in typical ODD scenarios. Moreover, driver distraction played a decisive role in 20% of the crashes considered in the analysis. Keeping these flawed behaviors out of the driving performances seems valuable on the basis of these percentages. Clearly, the analysis could be affected by the nature of the dataset considered, not intended for this type of evaluation. However, to prevent excessive uncertainties, each category of causes/fostering elements was analyzed on the basis of the type of crash. Where the pre-crash scenario was not clear, a conservative approach was adopted and no accident was considered as probably avoidable through the use of the Highway Chauffeur system. Besides, the analysis of the pre-crash scenario has already been employed for a first safety assessment of technologies not yet implemented on public roads [6, 14, 15, 21].

The estimated magnitude obtainable on safety is well represented by the total percentage of the addressed crashes (equal to 4,248), against the 6,408 considered in the ODD. The analysis highlighted that if each of the crashes occurred in 2016 had involved an HC vehicle, 66% of them could have been prevented or mitigated (on the basis of recorded causes and fostering elements). This represents the absolute number of crashes that, according to the analysis conducted, resulted to be very likely avoidable by an HC vehicle. In order to obtain an overall realistic estimation, this number must be projected considering the 10% market penetration assumed. Indeed, the greater number of crashes in the short term will still involve only traditional vehicles. Therefore, the estimated minimum number of crashes addressed through a 10% use of the Highway Chauffeur function on the Italian highways is equal to 6.6%. This represents a lower limit that does not take into account the crashes for which the dynamics could not be determined by ISTAT's public database. It must also be highlighted that the majority of crashes in the scenario of a 10% market penetration would not involve an HC vehicle. Therefore, both results should be considered, that is, 66 and 6.6%, each one with its statistical value.

To bluntly explicit the results obtained, if an effectiveness of 100% is considered, the owner of an HC vehicle would incur in at least 66% less crashes because the system can face successfully 66% of dangerous events when in the ODD. A traditional vehicle owner, in this scenario, would benefit from a crash reduction of 6.6%, which is the percentage of crashes avoidable because the other vehicle involved is an HC vehicle.

Besides, it is interesting to note that the 66% result referring to the crashes addressable by an HC vehicle is similar to the 60% result obtained in [14], representing the crashes addressable by an autonomous vehicle (Fig. 7.5).

Future works could concern the completeness and inclusiveness of the data used for the evaluation, either accessing ISTAT's limited-access database or employing a totally different dataset in order to validate, verify or tune the results presented in this article. Besides, works capable of considering also the accident-related events occurred in adverse weather conditions could improve and enhance the output of this analysis. Different approaches could also be explored to determine the number of crashes ascribable to the typical ODD of a Highway Chauffeur system. For example, one approach could be to consider the field tests related to the statistical value of the activation time and/or km driven in automated mode derived and applied on specific highways. Another approach could involve the adoption of effectiveness assumptions, such as the ones reported in [17, 20, 21] in order to avoid the 100% effectiveness hypothesis and obtain results closer to real outputs than to magnitude estimations. These coefficients should be obtained from field tests of vehicles achieving at least a level of automation equal to L2 or above, in order to consider the contribution of the sensors' suite and of software processing to the overall number

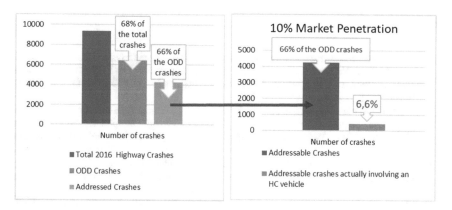

Fig. 7.5 Overall potential impact

of addressable crashes. In this paper, such approach was avoided because the field tests involving L3 systems on the Italian highways are not sufficient, therefore no effectiveness estimations could be found in literature. A step toward these results will be made within the C-Roads activities, during which HC vehicles will drive on the Italian highways and useful outputs will be obtained [22]. Besides, the performed analysis could be a useful input for cost–benefit analyses of the system in the short term. In this starting phase, in fact, there is the need to understand what the related benefit will be even though the implementation is not mature enough to use field tests data (both the prevention and the lowering of the consequences of a crash can be monetized and translated into economic benefits). As mentioned, the contribution of this paper indicates the magnitude of the safety impact obtainable from the use of the Highway Chauffeur system on the Italian highways. Therefore, future works in the same direction would be truly valuable to validate and tune the values obtained. Besides, the same analysis carried out on the Italian freeways would be valuable for an assessment in the medium term, in which it is foreseeable that also this type of road will be driven by a conditionally automated vehicle, as long as the ODD is met (the number of accidents on freeways in 2016 was equal to around 27,000 [23], against 9,360 on highways recorded by ISTAT). This kind of outputs could strongly improve the business case of an L3 system, even though not being exact results but rather estimations aimed at providing an order of magnitude. It is in fact clear that to state that a crash is avoidable on the basis of records should always be done with proper cautiousness, specifying that the objective of the work is not to determine what could have been avoided in the past, but rather to use the knowledge of the past to determine what will most likely be avoidable or at least addressable in the future.

References

1. NHTSA and US Department of Transportation: Automated driving systems 2.0—A vision for safety. NHTSA National Highway Traffic Safety Administration, DOT HS 812 442, September, 2017. Retrieved from https://www.nhtsa.gov/manufacturers/automated-driving-systems, 10 April 2018
2. ERTRAC Working Group Connectivity and Automated Driving: Automated driving roadmap status: final for publication v.7.0.. ERTRAC European Road Transport Research Advisory Council, May 2017. Retrieved from http://www.ertrac.org/index.php?page=ertrac-roadmap, 18 Jan 2018
3. Pendleton, S.D., Andersen, H., Du, X., Shen, X., Meghjani, M., Eng, H.Y., Rus, D., Ang, H.M.: Perception, planning, control, and coordination for autonomous vehicles. Machines 5(1), 6 (2017)
4. C-Roads Platform, Working Group 2 Technical Aspects, Taskforce 2 Service Harmonization. Common C-ITS Service Definitions Other Hazardous Locations Notification (OHLN), version 1.07, February 7, 2018
5. ISTAT. http://www4.istat.it/it/archivio/87539
6. Rau, P., Yanagisawa, M., Najm, W.G.: Target crash population of automated vehicles. In: 24th International Technical Conference on the Enhanced Safety of Vehicles (ESV), No. 15-0430, pp. 1–11. National Highway Traffic Safety Administration, Washington, DC (2015)
7. Centre for Connected and Autonomous Vehicles: Market forecast for connected and autonomous vehicles. Transport Systems Catapult, Milton Keynes, MK, United Kingdom. July, 2017. Retrieved from https://www.gov.uk/government/publications/connected-and-autonomous-vehicles-market-forcecast, 10 May 2018
8. AdaptIVe Project and A. Etemad: Deliverable D1.0 Final project result, June, 2017. Retrieved from https://www.adaptive-ip.eu/index.php/deliverables_papers.html, 17 Dec 2017
9. Suh, J., Kim, B., Yi, K.: Design and evaluation of a driving mode decision algorithm for automated driving vehicle on a motorway. IFAC-PapersOnLine 49(11), 115–120 (2016)
10. Katrakazas, C., Quddus, M., Chen, W., Deka, L.: Real-time motion planning methods for autonomous on-road driving: State-of-the-art and future research directions. Transp. Res. Part C: Emerg. Technol. 60, 416–442 (2015)
11. Best, A., Narang, S., Pasqualin, L., Barber, D., Manocha, D.: AutonoVi: autonomous vehicle planning with dynamic maneuvers and traffic constraints. In: IEEE/RSJ International Conference on Intelligent Robots and Systems (IROS). Vancouver, BC, Canada 24–28 Sept 2017
12. Jamson, A.H., Merat, N., Carsten, O.M.J., Lai, F.C.H.: Behavioural changes in drivers experiencing highly-automated vehicle control in varying traffic conditions. Transp. Res. Part C: Emerg. Technol. 30, 116–125 (2013)
13. Mcgehee, D.V., Brewer, M., Schwarz, C., Smith, B.W.: Review of automated vehicle technology: policy and implementation implications. In: IOWA Department of Transportation, DOT F 1700.7 (8–72), March 2016. https://rosap.ntl.bts.gov/view/dot/30702/dot_30702_DS1.pdf?
14. Eccles, K., Gross, F., Liu, M., Council, F.: Crash data analyses for vehicle-to-infrastructure communications for safety applications. US Department of Transportation, Report No. FHWA-HRT-11-040, November 2012
15. Najm, G.W., Koopmann, J., Smith, D.J., Brewer, J.: Frequency of target crashes for intellidrive safety systems. US Department of Transportation, October 2010
16. Kulmala, R.: Ex-ante assessment of the safety effects of intelligent transport systems. Accid. Anal. Prev. 42, 1359–1369 (2010)
17. Yue, L., Abdel-Aty, M., Wu, Y., Wang, L.: Assessment of the safety benefits of vehicles advanced driver assistance, connectivity and low level automation systems. Accid. Anal. Prev. 117, 55–64 (2018)
18. Huisman, M.: Impacts of (cooperative) adaptive cruise control on traffic flow—a simulation case study on the effects of (cooperative) adaptive cruise control on the A15 highway. Delft University of Technology, 2016

19. Hirz, M., Walzel, B.: Sensor and object recognition technologies for self-driving cars. Comput.-Aided Des. Appl. **15**(4), 501–508 (2018)
20. Li, T., Kockelman, K.M.: Valuing the safety benefits of connected and automated vehicle technologies. In: Transportation Research Board 95th Annual Meeting, 2016
21. Center for Transportation Research: Implications of connected and automated vehicles on the safety and operations of roadway networks: a final report. US Department of Transportation, Report No. FHWA/TX-16/0-6849-1, October 2016
22. Agriesti, S., Studer, L., Marchionni, G., Ponti, M., Gandini, P., Bresciani, C.: Highway chauffeur: state of the art and future evaluations—implementation scenarios and impact assessment. In: International Conference of Electrical and Electronic Technologies for Automotive, 2018
23. Li, T., Kockelman, K.M.: Valuing the safety benefits of connected and automated vehicle technologies. In: Transportation Research Board 95th Annual Meeting, 2016

Chapter 8
Recoverable Robustness Considering Carbon Tax in Weekly Berth and Quay Crane Planning

Qian Sun, Lu Zhen, Liyang Xiao and Zheyi Tan

Abstract The International Maritime Organization (IMO) has proposed to impose the carbon taxation on ports in the long term. The implementation of the carbon taxation policy will increase the operating cost and decrease its efficiency. Therefore, there is a trade-off between improving port operational efficiency and reducing carbon emission in ports. This study investigates the berth and quay cranes assignment planning, which aims at minimizing the total delay completion cost of all tasks, the total operating cost of quay cranes and the carbon taxation cost. In order to reflect the reality, some important uncertain factors are also considered. A recoverable robustness berth and quay crane assignment planning model is proposed. Numerical experiments are performed to demonstrate the applicability of the proposed model.

8.1 Introduction

Under the requirements of energy conservation and emission reduction, the carbon dioxide emission of the port area is highly concerned by governments and port operators. The International Maritime Organization (IMO) proposed to impose carbon emission tax on ports in the long term [1]. This policy will significantly increase the operating cost of ports. In the port operations, various container handling equipment is the main source of the carbon emission, such as quay cranes (QCs) [2]. In general, the allocation of QCs depends on the arrival time and workload of each vessel [3]. The former parameter directly determines vessel's dwelling time and the latter one represents the number of containers which are in demand of uploading and loading. We usually use the required QC-hours to reflect the number of containers. It is worth mentioning that the information of arrival time and workload will be notified to the port operator in advance. However, with the growth in container handling volumes, the two parameters mentioned above are usually inaccurate. Observations from ter-

Q. Sun (✉) · L. Zhen · L. Xiao · Z. Tan
School of Management, Shanghai University, Shanghai, China
e-mail: sunnnn@shu.edu.cn

L. Zhen
e-mail: lzhen@shu.edu.cn

© Springer Nature Singapore Pte Ltd. 2019
X. Qu et al. (eds.), *Smart Transportation Systems 2019*, Smart Innovation, Systems and Technologies 149, https://doi.org/10.1007/978-981-13-8683-1_8

minal operations point out that the exact arrival time of each vessel and the number of containers (i.e., workload) are unknown by the time when the weekly berth and quay crane planning is made [4].

Therefore, in order to more realistically reflect the trade-off between reducing carbon emissions and improving the efficiency of port operations, port operators need to optimize Berth and Quay Crane Assignment Planning (BACAP) considering uncertainties [5].

Reviewing literature, the studies on the carbon emission and energy-saving issue of apron-side operations at a container terminal are quite limited. To our best knowledge, there are only two related studies on QCs carbon emissions. Geerlings et al. [6] analyzed and calculated carbon emissions from all containers handling equipment at the Rotterdam Port Container Terminal. Wang et al. [7] proposed different carbon emission tax policies for BACAP. On the other hand, most of the papers about the integrated berth and quay crane planning have studied deterministic problems. Zeng et al. [8] avoided interruption through the rearrangement of QCs and the reallocation of ship berths. The authors assumed that the QC assignment does not change with time. A solution method based on Tabu search was proposed. Zhen et al. [9] designed a two-stage model to solve the BAP problem with uncertain arrival and processing time.

The rest of this paper is organized as follows. Section 8.2 describes the problem of BACAP in uncertain environment. Section 8.3 formulates a recoverable robust optimization model. Section 8.4 conducts numerical experiments to verify the effectiveness of the proposed model. Section 8.5 summarizes this study.

8.2 Problem Description

To better illustrate our problem, some important notations, i.e., QC-profile, taxation rate function, scenario, and so on, are explained as follows.

8.2.1 QC-Profile-Based QC Assignment Decision

To simplify the QC assignment decision in our problem, we follow the concept of QC-profile proposed by Giallombardo et al. [10]. According to the expected container handling workload, port operators will generate a set of QC-profile for incoming ships [11]. Two important parameters are defined for each QC-profile [12]. One parameter is the handling time by using QC-profile p for vessel i, denoted by h_{ip}. The other parameter is the number of QCs utilized in the mth time step if QC-profile p is assigned to vessel i, denoted by q_{ipm}.

8.2.2 Berth Allocation and QC Assignment

Berths are discrete areas of equal size. The purpose of this problem is to obtain a berth of each vessel, berthing start and end time, the QCs assignment problem (i.e., the QC-profile assignment) and buffer time as the baseline planning [13, 14]. At the same time, each scenario will be solved.

8.2.3 Taxation Rate Function

We consider a taxation rate function for carbon emission. The rate function is a constant function representing a unitary rate for any amount of carbon emission. Define $C_0^{CT}(Z) := r_0 Z$ as the tax function, where r_0 is the unitary taxation rate and Z is the amount of carbon emission [7].

8.2.4 Scenarios

There are two uncertain parameters for the problem. These are the vessel arrival time a_i^e and QC-hours requirement m_i for each vessel. If scenario ξ realizes, the arrival time $a_{i\xi}^e$ and the required QC-hours $m_{i\xi}$ will be applied for each vessel. We denote the combination of a arrival time step (i.e., $a_{i\xi}^e$) and a dwelling time (i.e., $m_{i\xi}$) as a scenario of each vessel.

8.3 Recoverable Robustness Model

The integrated model of BACAP aims to minimize the service cost in baseline schedule and each scenario, which is related to the deviation from vessels' expected turnaround time windows, QC operation costs and the carbon taxation costs [15]. The model also aims to the adjustment cost from baseline schedule to scenario, to be more specific, the part is linked to the recovery cost from baseline schedule for each scenario. All of these costs are multiplied with the realization probability of the given scenario. The integer variable $v_{i\xi}$ shows how much vessel i is postponed due to the recovery for scenario ξ. While $l_{i\xi}$ corresponds to how late operations go beyond the end of buffer time (θ_{it}) for vessel i.

Indices and sets:

i, j indices of vessels;
V set of vessels;
b index of berths;
B set of berths;

p index of QC-profiles;

P_i set of QC-profiles for vessel i, $i \in V$;

t index of time steps;

T set of time steps;

ξ index of scenarios;

Φ set of scenarios.

Parameters:

M a large positive number;

m_i expected workload in QC-hours for vessel i in baseline schedule, $i \in V$;

$m_{i\xi}$ actual workload in QC-hours for vessel i in scenario ξ, $i \in V, \xi \in \Phi$;

$a_{i\xi}^e$ actual service time steps for vessel i in scenario ξ, $i \in V, \xi \in \Phi$;

h_{ip} handling time of vessel i by using QC-profile p (unit: time step), $i \in V, p \in P_i$;

q_{ipm} number of QCs used by using QC-profile p at the mth time step, $p \in P_i, i \in V, m \in \{1, \ldots, h_{ip}\}$;

Q maximum number of QCs available in each time step;

C_i unit penalty cost (for vessel i, $i \in V$) assigned to the earliness and tardiness beyond the vessel's expected service time window (unit: \$/hr);

C^2 Cost rate per QC-hour of operation;

$\rho(\xi)$ Probability of realization for scenario ξ, $\xi \in \Phi$;

CO Carbon emission factor for a QC (unit: kg/kWh);

$[a_i^e, b_i^e]$ expected service time window for vessel i, $i \in V$.

Variables:

w_{ib} binary variable, set to one if berth b is allocated to vessel i in baseline schedule and to zero otherwise, $i \in V, b \in B$;

$w_{ib\xi}$ binary variable, set to one if berth b is allocated to vessel i in scenario ξ and to zero otherwise, $i \in V, b \in B, \xi \in \phi$;

δ_{ijb} binary variable, set to one if vessel i, j dwell at berth b, and vessel i dwells before vessel j in baseline schedule and to zero otherwise, $i, j \in V, i \neq j, b \in B$;

$\delta_{ijb\xi}$ binary variable, set to one if both vessel i, j dwell at berth b, and vessel i dwells at the berth before vessel j in scenario ξ and to zero otherwise, $i, j \in V, i \neq j, b \in B, \xi \in \phi$;

γ_{ip} binary variable, set to one if vessel i is served by QC-profile p in baseline schedule and to zero otherwise, $i \in V, p \in P_i$;

$\gamma_{ip\xi}$ binary variable, set to one if vessel i is served by QC-profile p in scenario ξ and to zero otherwise, $i \in V, p \in P_i, \xi \in \phi$;

σ_t integer, the number of used QCs at time step t in baseline schedule, $t \in T$;

$\sigma_{t\xi}$ integer, the number of used QCs at time step t in scenario $\xi, t \in T, \xi \in \phi$;

μ_{it} binary variable, set to one if vessel i begins handling in time step t in baseline schedule and to zero otherwise, $i \in V, t \in T$;

$\mu_{it\xi}$ binary variable, set to one if vessel i begins handling in time step t in scenario ξ and to zero otherwise, $i \in V, t \in T, \xi \in \phi$;

η_{ipt} binary variable, set to one if vessel i is served by p and begins handling in time step t in baseline schedule and to zero otherwise, $i \in V, t \in T, p \in P_i$;

$\eta_{ipt\xi}$ binary variable, set to one if vessel i is served by p and begins handling in time step t in scenario ξ and to zero otherwise, $i \in V, t \in T, p \in P_i, \xi \in \phi$;

α_i integer, the start time step of the handling for vessel i, $i \in V$;

$\alpha_{i\xi}$ integer, the start time step of the handling for i in scenario ξ, $i \in V, \xi \in \phi$;

β_i integer, the depart time of the handling for vessel i, $i \in V$;

$\beta_{i\xi}$ integer, the depart time of the handling for vessel i in scenario ξ, $i \in V, \xi \in \phi$;

θ_{it} binary variable, set to one if the reserved berthing time window for vessel i ends at time step t in baseline schedule and to zero otherwise, $i \in V, t \in T$;

$\upsilon_{i\xi}$ integer, relative lateness of vessel i with respect to baseline schedule for scenario ξ, $i \in V, \xi \in \phi$;

$\iota_{i\xi}$ integer, duration of handling time beyond the buffer time for vessel i in scenario ξ, $i \in V, \xi \in \phi$;

Z_ξ the quantity of carbon emission (tons) in scenario ξ, $\xi \in \phi$.

Mathematical model:

$$min \begin{cases} \sum_{i\in V} C_i\left[(\alpha_i - a_i^e)^+ + (\beta_i - b_i^e)^+\right] + C^2 \sum_{t\in T} \sigma_t \\ + \sum_{i\in \upsilon} \sum_{\xi\in\phi} C_i\rho(\xi)(\upsilon_{i\xi} + \iota_{i\xi}) + \\ \sum_{i\in V} \sum_{\xi\in\phi} C_i\rho(\xi)\left[(\alpha_{i\xi} - a_{i\xi}^e)^+ + (\beta_{i\xi} - b_i^e)^+\right] + \\ C^2 \sum_{t\in T} \sum_{\xi\in\phi} \rho(\xi)\sigma_{t\xi} + \sum_{\xi\in\phi} \rho(\xi)C_0^{CT}(Z_\xi) \end{cases} \tag{8.1}$$

$$s.t. \quad \sum_{b\in B} \omega_{ib} = 1 \quad \forall i \in V \tag{8.2}$$

$$\sum_{t\in T} \mu_{it} = 1 \quad \forall i \in V \tag{8.3}$$

$$\sum_{t\in T} \mu_{it}t = \alpha_i \quad \forall i \in V \tag{8.4}$$

$$\alpha_i + \sum_{p\in P_i} \gamma_{ip}h_{ip} - 1 = \beta_i \quad \forall i \in V \tag{8.5}$$

$$\sum_{t\in\{min_{p\in P_i}\{h_{ip}\},H\}} \theta_{it} = 1 \quad \forall i \in V \tag{8.6}$$

$$\beta_i - t \le (1 - \theta_{it})M \quad \forall i \in V, t \in T \tag{8.7}$$

$$\alpha_j + (1 - \delta_{ijb})M \ge \sum_{t\in T} \theta_{it}t \quad \forall i, j \in V, i \ne j, b \in B \tag{8.8}$$

$$\delta_{ijb} + \delta_{jib} \le \omega_{ib} \quad \forall i, j \in V, i \ne j, b \in B \tag{8.9}$$

$$\delta_{ijb} + \delta_{jib} \ge \omega_{ib} + \omega_{jb} - 1 \quad \forall i, j \in V, i \ne j, b \in B \tag{8.10}$$

$$\sum_{p \in P_i} \gamma_{ip} = 1 \quad \forall i \in V \tag{8.11}$$

$$\eta_{ipt} \geq \gamma_{ip} + \mu_{it} - 1 \quad \forall i \in V, p \in P_i, t \in T \tag{8.12}$$

$$\sigma_t = \sum_{i \in V} \sum_{p \in P_i} \sum_{m=max\{1,t-h_{ip}+1\}}^{t} \eta_{ipm} q_{ip(t-m+1)} \quad \forall t \in T \tag{8.13}$$

$$\sigma_t \leq Q \quad \forall t \in T \tag{8.14}$$

$$\sum_{b \in B} \omega_{ib\xi} = 1 \quad \forall i \in V, \xi \in \phi \tag{8.15}$$

$$\sum_{t \in T} \mu_{it\xi} = 1 \quad \forall i \in V, \xi \in \phi \tag{8.16}$$

$$\sum_{t \in T} \mu_{it\xi} t = \alpha_{i\xi} \quad \forall i \in V, \xi \in \phi \tag{8.17}$$

$$\alpha_{i\xi} + \sum_{p \in P_i} \gamma_{ip\xi} h_{ip} - 1 = \beta_{i\xi} \quad \forall i \in V, \xi \in \phi \tag{8.18}$$

$$\alpha_{j\xi} + \left(1 - \delta_{ijb\xi}\right) M \geq \beta_{i\xi} \quad \forall i, j \in V, i \neq j, b \in B, \xi \in \phi \tag{8.19}$$

$$\delta_{ijb\xi} + \delta_{jib\xi} \leq \omega_{ib\xi} \quad \forall i, j \in V, i \neq j, b \in B, \xi \in \phi \tag{8.20}$$

$$\delta_{ijb\xi} + \delta_{jib\xi} \geq \omega_{ib\xi} + \omega_{jb\xi} - 1 \quad \forall i, j \in V, i \neq j, b \in B, \xi \in \phi \tag{8.21}$$

$$\sum_{p \in P_i} \gamma_{ip\xi} = 1 \quad \forall i \in V, \xi \in \phi \tag{8.22}$$

$$\eta_{ipt\xi} \geq \gamma_{ip\xi} + \mu_{it\xi} - 1 \quad \forall i \in V, p \in P_i, t \in T, \xi \in \phi \tag{8.23}$$

$$\sigma_{t\xi} = \sum_{i \in V} \sum_{p \in P_i} \sum_{m=max\{1,t-h_{ip}+1\}}^{t} \eta_{ipm\xi} q_{ip(t-m+1)} \quad \forall t \in T, \xi \in \phi \tag{8.24}$$

$$\sigma_{t\xi} \leq Q \quad \forall t \in T, \xi \in \phi \tag{8.25}$$

$$\upsilon_{i\xi} \geq \beta_{i\xi} - \beta_i - \left(a_{i\xi}^e - a_i^e\right) \quad \forall i \in V, \xi \in \phi \tag{8.26}$$

$$\iota_{i\xi} \geq \beta_{i\xi} - \sum_{t \in T} \theta_{it} t \quad \forall i \in V, \xi \in \phi \tag{8.27}$$

$$Z_\xi = \sum_{t \in T} \sigma_{t\xi} \times co \tag{8.28}$$

$$w_{ib}, \delta_{ijb}, \gamma_{ip}, \mu_{it}, \eta_{ipt}, \theta_{it} \in \{0, 1\} \quad \forall i, j \in V, t \in T, b \in B, p \in P_i \tag{8.29}$$

$$w_{ib\xi}, \delta_{ijb\xi}, \gamma_{ip\xi}, \mu_{it\xi}, \eta_{ipt\xi} \in \{0, 1\} \quad \forall i, j \in V, t \in T, b \in B, p \in P_i, \xi \in \phi \tag{8.30}$$

$$\sigma_t, \sigma_{t\xi}, \alpha_i, \alpha_{i\xi}, \beta_i, \beta_{i\xi}, \upsilon_{i\xi}, \iota_{i\xi}, Z_\xi \geq 0 \quad \forall i \in V, t \in T, \xi \in \phi \tag{8.31}$$

Constraints (8.2) mean that each vessel can be allocated to only one berth. Constraints (8.3) state that each vessel starts handling in a certain time step. Constraints (8.4) connect the two handling start time decision variables (i.e., μ_{it} and α_i). If vessel i begins handling in time step t (i.e., $\mu_{it} = 1$), the start time step of the handling for vessel i is time step t. Constraints (8.5) link the start time step and the end time step of the vessels. Constraints (8.6) ensure that there is only one period for the ending of the buffer time for each vessel. Constraints (8.7) set the ending of the buffer times correctly. The buffer ending time (θ_{it}) should be greater than or equal to the operation end time (β_i). Constraints (8.8) ensure that the vessels cannot overlap considering their time windows. The berthing time window for vessel i is from α_i to θ_{it}. Constraints (8.9) and Constraints (8.10) guarantee that if two vessels are allocated to the same berth, there must be a time sequence for the two vessels dwelling at the berth. Constraints (8.11) stipulate that only one QC-profile is assigned to each vessel. Constraints (8.12) link two decision variables η_{ipt} and μ_{it}, which are both related to the start time of handling. Constraints (8.13) calculate the number of QCs used in each time step. Constraints (8.14) guarantee that the number of QCs used in each time step cannot exceed the capacity Q. Constraints (8.15–8.18) and Constraints (8.20–8.25) are copies of Constraints (8.2–8.5) and Constraints (8.9–8.14) in each scenario. Constraints (8.26) set $\upsilon_{i\xi}$ variables. The term $\beta_{i\xi} - \beta_i$ at right-hand side corresponds to how late vessel i is in scenario ξ solution compared to baseline schedule. There could be the case that vessel i itself is late in scenario ξ so that it is meaningless to add the cost $\beta_{i\xi} - \beta_i$ to objective function directly. For this reason, we subtract the lateness from expected arrival $\alpha_{i\xi} - \alpha_i$ from the lateness in the operations. Constraints (8.27) set $\iota_{i\xi}$ variable which corresponds to lateness compared to the end of the buffer times for vessel i in scenario ξ. The positive difference between $\beta_{i\xi}$ and $\sum_{t \in T} \theta_{it}$ reflects the lateness in the finishing time for vessel i in scenario ξ. The bigger values of this variable mean higher rescheduling efforts in the yard and the quay area. Constraints (8.28) give the quantity of carbon emission in each scenario. Constraints (8.29–8.31) define the domain of decision variables.

8.4 Computational Experiments

The experiments were run on a PC equipped with a 3.2 GHz Intel Core i5 CPU and 16 GB RAM. All of the algorithms were programmed in C# (VS2015), and the model was solved by CPLEX 12.5. The time limit for all test instances was 3 h (10,800 s).

Table 8.1 Vessel types and characteristics information

Vessel		QC-profile specifications			
Class	Proportion	Range of used QCs	Range of handling time	Range of workload	Range of workload in scenarios
Feeder	1/3	1–3	2–4	2–5	2–7
Medium	1/3	2–4	3–5	6–14	6–17
Jumbo	1/3	3–5	4–6	15–20	13–21

8.4.1 Generation of Test Cases

We generate 15 instances with different characteristics. The vessels are distinguished into three classes, with different technical specifications (see Table 8.1). The planning period is one week. The planning period has 42 time steps. The number of discrete scenarios (S) is 2 and 5. The realization probability ($\rho(\xi)$) for each scenario ξ is equal for one instance, and it depends on the number of scenarios (0.6/S). The cost of operating QC is set as $C^2 = 100\$/hr$ [7]. The energy consumption of QCs is set to CO $= 149.7$ KWh/hr $\times 1.0935$ kg/KWh [16].

8.4.2 Evaluating the Effectiveness of the Proposed Models

The results of the comparative experiments the S $= 2$ and S $= 5$ are listed as follows (Table 8.2).

Table 8.2 shows the solution method that solves the problem directly by CPLEX, which provides the optimal solution. The experimental results show that the model is effective, but because the problem is NP-hard, the result of solving the large-scale problem CPLEX is not ideal, even in the case of 60 ships in 5 scenarios, it cannot be solved.

8.5 Conclusions

This study proposes a berth and quay cranes assignment problem, which optimizes the operational costs of ports. Under the requirement of green port, we take carbon tax as part of the objective function. In addition, some uncertain factors are also considered. The problem is formulated as a recoverable robustness model and then solved by commercial solver CPLEX. As can be observed from the experimental results, the computation time increases dramatically as the problem size scales up. Thus, future research will focus on efficient solution method to solve the problem in reasonable computation time.

Table 8.2 Solution of solving the model by CPLEX solver

#	V	B	Q	S = 2			S = 5		
				LB	G%	t(s)	LB	G%	t(s)
1-1	21	3	7	20749.84	0.00	94	21925.74	0.00	77
1-2	21	3	7	26831.98	0.01	75	21918.93	0.00	120
1-3	21	3	7	24463.78	0.00	44	32443.66	0.01	100
2-1	30	4	11	29519.04	0.00	122	33549.59	0.00	195
2-2	30	4	11	29519.04	0.00	121	41298.70	0.00	960
2-3	30	4	11	33272.1	0.00	120	43459.43	0.01	5880
3-1	36	5	12	47302.76	0.03	7200	37863.02	0.01	2880
3-2	36	5	12	42049.67	0.05	7200	39249.91	0.20	10800
3-3	36	5	12	37855.02	0.01	4620	45690.25	1.20	10800
4-1	48	7	18	47240.84	0.00	672	56992.78	0.00	1380
4-2	48	7	18	54841.61	0.00	2660	56990.38	0.00	2244
4-3	48	7	18	47240.84	0.00	573	53683.21	0.10	10800
5-1	60	8	21	62787.53	1.40	10800	–	–	–
5-2	60	8	21	68348.35	1.70	10800	–	–	–
5-3	60	8	21	66509.17	1.40	10800	–	–	–

References

1. Wang, S., Zhen, L., Zhuge, D.: Dynamic programming algorithms for selection of waste disposal ports in cruise shipping. Transp. Res. Part B Methodol. **108**, 235–248 (2018)
2. Wang, S., Qu, X., Yang, Y.: Estimation of the perceived value of transit time for containerized cargoes. Transp. Res. Part A Policy Pract. **78**, 298–308 (2015)
3. Lee, C.-Y., Song, D.-P.: Ocean container transport in global supply chains: overview and research opportunities. Transp. Res. Part B Methodol. **95**, 442–474 (2017)
4. Zhen, L., Wang, K.: A stochastic programming model for multi-product oriented multi-channel component replenishment. Comput. Oper. Res. **60**, 79–90 (2015)
5. Zhen, L.: Tactical berth allocation under uncertainty. Eur. J. Oper. Res. **247**(3), 928–944 (2015)
6. Geerlings, H., van Duin, R.: A new method for assessing CO_2-emissions from container terminals: a promising approach applied in Rotterdam. J. Clean. Prod. **19**(6), 657–666 (2011)
7. Wang, T., Wang, X., Meng, Q.: Joint berth allocation and quay crane assignment under different carbon taxation policies. Transp. Res. Part B Methodol. **117**, 18–36 (2018)
8. Zeng, Q., Yang, Z., Hu, X.: Disruption recovery model for berth and quay crane scheduling in container terminals. Eng. Optim. **43**(9), 967–983 (2011)
9. Zhen, L., Lee, L.H., Chew, E.P.: A decision model for berth allocation under uncertainty. Eur. J. Oper. Res. **212**(1), 54–68 (2011)
10. Giallombardo, G., et al.: Modeling and solving the tactical berth allocation problem. Transp. Res. Part B Methodol. **44**(2), 232–245 (2010)
11. Wang, K., et al.: Column generation for the integrated berth allocation, quay crane assignment, and yard assignment problem. Transp. Sci. **52**(4), 812–834 (2018)
12. Zhen, L., et al.: Daily berth planning in a tidal port with channel flow control. Transp. Res. Part B Methodol. **106**, 193–217 (2017)
13. Wang, S., Wang, X.: A polynomial-time algorithm for sailing speed optimization with containership resource sharing. Transp. Res. Part B Methodol. **93**, 394–405 (2016)

14. Shang, X.T., Cao, J.X., Ren, J.: A robust optimization approach to the integrated berth allocation and quay crane assignment problem. Transp. Res. Part E Logist. Transp. Rev. **94**, 44–65 (2016)
15. Iris, Ç., et al.: Integrated berth allocation and quay crane assignment problem: set partitioning models and computational results. Transp. Res. Part E Logist. Transp. Rev. **81**, 75–97 (2015)
16. Chang, D., et al.: Integrating berth allocation and quay crane assignments. Transp. Res. Part E Logist. Transp. Rev. **46**(6), 975–990 (2010)

Chapter 9
Research on Freeway Mainline Fuzzy Logic Control Based on Dynamic Traffic Evaluation

Sheng Zhao, Weiming Liu, Huiying Wen and Weiwei Qi

Abstract For now, traffic safety and traffic congestion are two difficult problems which freeway traffic management must deal with. And there are so many factors influencing highway traffic, in which some factors are characterized as dynamic, such as traffic stream characteristics, weather conditions, road environmental illumination, and so on. Considering weather conditions, road conditions, and traffic conditions, an evaluation index system is established. And based on the index system, real-time traffic safety level and traffic congestion level are evaluated using fuzzy theory and neural network. Then the highway mainline fuzzy logic control method is proposed based on dynamic traffic evaluation results, which will improve the safety of freeway traffic and alleviate the traffic congestion of freeway through balancing traffic flow on freeway.

9.1 Introduction

Freeway had been originally conceived to provide convenient and efficient mobility to road users. However, with the continuous increase of car ownership and demand, recurrent and non-recurrent freeway congestions have become more and more frequent [1]. Traffic congestions have negative influences on travel times, traffic safety, fuel consumption, and environmental pollution [2, 3]. In order to alleviate traffic congestion and improve traffic safety, many freeway traffic control methods are proposed. Andreas Hegyi et al. came up with a method of acquiring optimal speed limit based on METANET traffic model [4]. Rongjie Yu et al. proposed a control model based on collision risk assessment to get an optimal speed limit with low collision risk [5]. Tang-Hsien Chang et al. constructed an innovative ramp-metering control model to optimize mainline traffic by adjusting metering rates [6]. Zhibin Li et al. presented optimal variable speed limit control model to decrease collision risks and injury severity by real-time crash risk and severity prediction [7].

S. Zhao · W. Liu · H. Wen · W. Qi (✉)
School of Civil Engineering and Transportation, South China University of Technology, Guangzhou 510641, China
e-mail: qwwhit@163.com

© Springer Nature Singapore Pte Ltd. 2019
X. Qu et al. (eds.), *Smart Transportation Systems 2019*, Smart Innovation, Systems and Technologies 149, https://doi.org/10.1007/978-981-13-8683-1_9

The main purpose of freeway mainline control has two aspects: one is to control traffic vehicle moving speed to accommodate driving environment to avoid traffic accidents and another is to equalize vehicle distribution on freeway to avoid traffic congestion caused by excessive traffic density and increase capacity of the freeway infrastructure. Traffic safety and traffic congestion are affected by weather conditions, road conditions, and traffic conditions in varying degrees. Meanwhile different weather conditions and traffic running conditions which are dynamic have different effects on freeway traffic. These dynamic affecting factors must be taken into consideration when making freeway traffic control strategies. But most present freeway traffic control methods are established based on analysis of the traffic flow characteristics without considering those factors [8–14]. Therefore, a kind of freeway mainline control method considering weather conditions, road conditions, and traffic conditions is proposed in this paper which realizes the control aim of freeway traffic effectively and safely.

9.2   Dynamic Evaluation of Freeway Traffic Based on Fuzzy Theory and Neural Network

Firstly, freeway traffic evaluation index system should be proposed. Freeway traffic evaluation index system contains two parts, which are indices for describing traffic conditions and indices for describing traffic environment. Indices for describing traffic running conditions include traffic flow, traffic density, and average travel speed. And indices for describing traffic environment include road curve radius, longitudinal gradient, grade length, adhesion coefficient of road surface, and visibility.

Dynamic evaluation of freeway traffic has two dimensions, one is traffic safety level evaluation and the other is traffic congestion level evaluation. Traffic congestion level can be evaluated using traffic density and vehicle moving speed. And traffic safety level is related to traffic environment, depending on what degree traffic vehicle moving speed accommodates driving environment. But the effects of driving environment on traffic safety, like road curve radius, longitudinal gradient, grade length, adhesion coefficient of road surface, and visibility, can't be quantified easily, because of their different dimension and influence mechanism. Therefore, the traffic environment and road conditions are evaluated using the fuzzy comprehensive assessment model. And then the evaluation result is used in dynamic evaluation of freeway traffic as an index.

9.2.1   Fuzzy Comprehensive Evaluation of Traffic Environment and Road Conditions

The fuzzy comprehensive assessment of traffic environment and road conditions adopts these indices, such as road curve radius, longitudinal gradient, grade length, adhesion coefficient of road surface and visibility, and so on. The steps to evaluate driving environment are as follows:

Step one: Establish evaluation factor set U.

U = {U1, U2, U3, U4, U5} = {road curve radius, longitudinal gradient, grade length, adhesion coefficient of road surface, visibility}.

Step two: Establish weighted values set A. Every index has different effect on assessment result, it is necessary to assign different weighted value. The weighting matrix is A = {0.10, 0.8, 0.06, 0.38, 0.38}.

Step three: Establish fuzzy evaluation result set V. The fuzzy assessment result set of every evaluation factor corresponds to V1 = {0.1, 0.3, 0.5, 0.7, 0.9}, V2 = {0.9, 0.7, 0.5, 0.3, 0.1}, V3 = {0.9, 0.7, 0.5, 0.3, 0.1}, V4 = {0.1, 0.3, 0.5, 0.7, 0.9}, V5 = {0.1, 0.3, 0.5, 0.7, 0.9}. Then the fuzzy assessment result set V is as formula (9.1).

$$V = [V1, V2, V3, V4, V5]^{T} = \begin{bmatrix} 0.1 & 0.3 & 0.5 & 0.7 & 0.9 \\ 0.9 & 0.7 & 0.5 & 0.3 & 0.1 \\ 0.9 & 0.7 & 0.5 & 0.3 & 0.1 \\ 0.1 & 0.3 & 0.5 & 0.7 & 0.9 \\ 0.1 & 0.3 & 0.5 & 0.7 & 0.9 \end{bmatrix} \qquad (9.1)$$

Step four: Establish fuzzy evaluation matrix R. The membership functions of every evaluation factor are as following figures (Figs. 9.1, 9.2, 9.3, 9.4, 9.5).

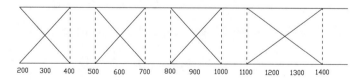

Fig. 9.1 Membership function curves of road curve radius (unit: m)

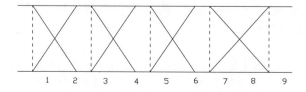

Fig. 9.2 Membership function curves of longitudinal gradient (unit: %)

Fig. 9.3 Membership function curves of grade length (unit: m)

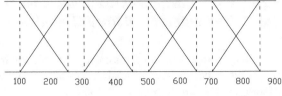

Fig. 9.4 Membership function curves of visibility (unit: m)

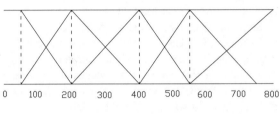

Fig. 9.5 Membership function curves of adhesion coefficient of road surface

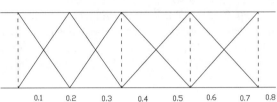

According to the membership functions curves above, we can get the fuzzy evaluation matrix R.

$$R = [R_1, R_2, R_3, R_4, R_5]^T = \begin{bmatrix} r_{11} & r_{12} & r_{13} & r_{14} & r_{15} \\ r_{21} & r_{22} & r_{23} & r_{24} & r_{25} \\ r_{31} & r_{32} & r_{33} & r_{34} & r_{35} \\ r_{41} & r_{42} & r_{43} & r_{44} & r_{45} \\ r_{51} & r_{52} & r_{53} & r_{54} & r_{55} \end{bmatrix} \quad (9.2)$$

Step five: The evaluation result G can be gotten through the following formula.

$$G = \left[V_1 R_1^T, V_2 R_2^T, V_3 R_3^T, V_4 R_4^T, V_5 R_5^T \right] A^T \quad (9.3)$$

9.2.2 Dynamic Evaluation of Freeway Traffic Based on BP Neural Network

As is mentioned above, dynamic evaluation of freeway traffic includes two sides. One is evaluation of traffic safety level. Another is evaluation of traffic congestion level. Whether the freeway traffic is safe or not is related to whether vehicle moving speed accommodates driving environment. The process of traffic safety level evaluation is as follows:

Step one: Establish evaluation indices. The assessing indices are vehicle moving speed, traffic density, and the evaluation result G of traffic environment and road conditions.

Step two: Design the input layer, hidden layer, and output layer of BP neural network [15]. The number of input neurons is 3. The number of output neurons is 1 and its threshold is between 0 and 1. And the bigger its value is, means the higher traffic safety level is. The number of hidden neurons is determined by the formula, $b = \sqrt{m+n} + \alpha$, in which m is the number of input neurons, n is number of output neurons, and α is constant between 1 and 10.

Step three: Training sample. In order to improve the convergence speed of sample training, it is necessary to normalize the input data using compressibility method. And then confirm input sample, test sample, learning rate, and the maximum number of iteration.

Evaluation of traffic congestion level based on BP neural network is similar to the evaluation of traffic safety level. What is different is the assessing indices are vehicle moving speed, traffic density, and that the number of input neurons is 2. The rest steps are same as evaluation of traffic safety level.

9.3   Freeway Mainline Fuzzy Logic Control Based on Dynamic Traffic Evaluation

The main emphasis of freeway fuzzy logic mainline control based on dynamic traffic evaluation is to make speed limit strategy according to the dynamic evaluation result of freeway traffic. The input variables of freeway fuzzy logic mainline control are evaluation result of traffic safety level (R1), vehicle moving speed (V), and evaluation result of traffic congestion level (R2). The sketch of freeway fuzzy logic mainline control is shown in Fig. 9.6.

The output variable of freeway fuzzy logic mainline control is speed limit of mainline (SL). Steps of freeway fuzzy logic mainline control are as follows:

Step one: Construct fuzzy set of input variables and output variable. Fuzzy set of input variables and output variables is described as very small, small, middle, big, and very big. Then the fuzzy sets of input variables are as follows:

{R1_VS, R1_S, R1_M, R1_B, R1_VB},
{R2_VS, R2_S, R2_M, R2_B, R2_VB},

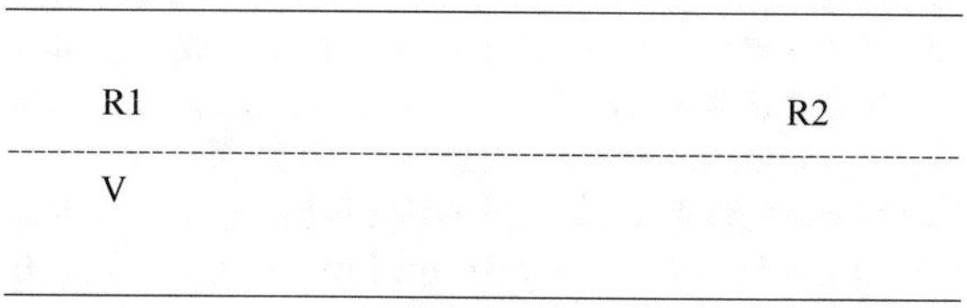

Fig. 9.6 Sketch of freeway fuzzy logic mainline

Fig. 9.7 Membership
function curves of input
variable R1

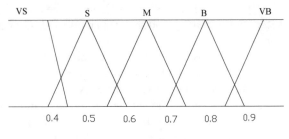

Fig. 9.8 Membership
function curves of input
variable R2

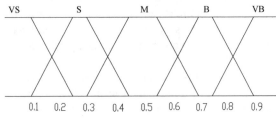

Fig. 9.9 Membership
function curves of input
variable V (unit: Km/h)

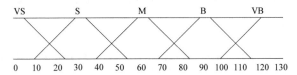

Fig. 9.10 Membership
function curves of output
variable SL (unit: Km/h)

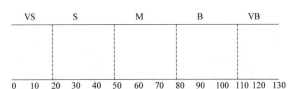

{V_VS, V_S, V_M, V_B, V_VB}.

The fuzzy set of output variable is as follows:

{SL_VS, SL_S, SL_M, SL_B, SL_VB}.

And the membership functions of input variables and output variable are shown in following figures (Figs. 9.7, 9.8, 9.9, 9.10).

Step two: Construct fuzzy rule-based system. The fuzzy logic controller has three input variables. And every variable has five ranks, then there will be 125 fuzzy control rules. But there is no need to limit the speed when vehicle moving speed is very small or vehicle moving speed matches traffic environment very well. Another situation is that it takes some fuzzy rule as long as input variables are above some value. For example, the control statement IF N_R1_B AND N_R2_S AND A_V_B THEN SL_M means when the rank of R1 is below R1_B, and the rank of R1 is below R2_S, then it carries out middle speed limit strategy. So, there are four fuzzy rules in the rule-based system.

(1)   IF N_R1_B AND V_VB THEN SL_B,
(2)   IF N_R1_B AND V_B THEN SL_M,
(3)   IF N_R1_B AND V_M THEN SL_S,
(4)   IF N_R1_B AND N_R2_S AND A_V_B THEN SL_M.

Step three: Construct fuzzier. The rule of fuzzier is when some input variables get two different values according to membership function curves, then assign the bigger one. If those two values are equal, then assign the worse one. For example, if R2 is 0.6, $f_M(R2)$ is 0.67 and $f_B(R2)$ is 0.33, then assign R2_M to f(R2). If R2 is 0.625, $f_M(R2)$ and $f_B(R2)$ are both 0.5, then assign R2_B to f(R2).

Step four: Construct canceller. A set of input variables may map several control rules. The role of canceller is to realize a set of input variables map only one control rule. The canceller is as follows:

$$SL = \frac{\sum\limits_{i=1}^{N} w_i c_i f_i / d_i}{\sum\limits_{i=1}^{N} w_i f_i / d_i} \tag{9.4}$$

SL: speed limit;
N: number of fuzzy rules;
w_i: weight of fuzzy rule number i;
c_i: center of output fuzzy set mapped by fuzzy rule number i;
d_i: width of output fuzzy set mapped by fuzzy rule number i;
f_i: fuzzy implication of fuzzy rule number i.

9.4   Conclusion

A dynamic freeway traffic evaluation method is proposed based on fuzzy theory and neural network, which includes two aspects including traffic safety level evaluation and traffic congestion level evaluation. And the freeway traffic evaluation index system consisted of adhesion coefficient of road surface, visibility, traffic density, and vehicle moving speed is established. And based on the dynamic freeway traffic evaluation results, a kind of freeway mainline fuzzy logic control model is built, which provides an effective method for freeway mainline control.

Acknowledgements The study is supported by the National Natural Science Foundation of China (NO. 51578247 & 71701070), the Natural Science Foundation of Guangdong Province (NO. 2016A030310427), and the Science and Technology Project of Guangzhou City (NO. 201804010466).

References

1. Carlson, R.C., Ioannis, P.: Optimal mainstream traffic flow control of large-scale motorway networks. Transp. Res. Part C: Emerg. Technol. **18**(6), 193–212 (2010).
2. Carlson, R.C., Ioannis, P.: Local feedback-based mainstream traffic flow control on motorways using variable speed limits. IEEE Trans. Intell. Transp. Syst. **12**(4), 1261–1276 (2011).
3. Kuang, Y., Qu, X., Wang, S.: A tree-structured crash surrogate measure for freeways. Accid. Anal. Prev. **77**, 137–148 (2015).
4. Andreas, H., Bart, D.S., Hans, H.: Model predictive control for optimal coordination of ramp metering and variable speed limits. Transp. Res. Part C **13**(3), 39–41 (2004)
5. Rongjie, Y., Mohamed, A.: An optimal variable speed limits system to ameliorate traffic safety risk. Transp. Res. Part C **46**, 235–246 (2014)
6. Tang, H., Chang, Z.L.: Optimization of mainline traffic via an adaptive co-ordinated ramp-metering control model with dynamic OD estimation. Transp. Res. Part C **10**(2), 99–120 (2002)
7. Zhibin, L., Pan, L., Chengcheng, X., Wei, W.: Optimal mainline variable speed limit control to improve safety on large-scale freeway segments. Computer-Aided Civil and Infrastructure Engineering **31**, 366–380 (2016)
8. Lu, K., Xu, J.: Design methods for main line speed control of expressway traffic flow density. Freeway **4**, 16–18 (2008)
9. Liang, X., Liu, Z., Mao, Z.: Fuzzy ramp control in freeway and simulation research. J. Syst. Simul. **17**(2), 444–447 (2005).
10. Kotsialos, A., Papageorgiou, M.: A hierarchical ramp metering control scheme for freeway networks. In: American Control Conference, pp. 2257–2262. Portland, OR, USA (2005)
11. Shih, C.L., Hsun, J.C.: Chaos and control of discrete dynamic traffic model. J. Frankl. Inst. (S0016–0032) **342**(7) 839–851 (2005).
12. Wang, Y.: Study of traffic congestion's simulation based on cellular automaton model. J. Syst. Simul. (S1004-731X) **22**(9), 2149- 2154 (2010).
13. Yang, Q., Ma, M., Liang, S., Li, Z.: Stair-like control strategies of variable speed limit for bottleneck regions on freeway. J. Southwest Jiaotong Univ. **50**(2), 354–360 (2015)
14. Bhourin, Haj, S., Kaupplaj.: Isolated versus coordinated ramp metering: field evaluation results of travel time reliability and traffic impact. Transp. Res. Part C: Emerg. Technol. **28**, 155–167 (2013).
15. Zhou, M., Qu, X., Li, X.: A recurrent neural network based microscopic car following model to predict traffic oscillation. Transp. Res. Part C **84**, 245–264 (2017)

Chapter 10
Time-Domain and Frequency-Domain Analysis of Driver's ECG Characteristics in Rainy Environment

Weiwei Qi, Zhuoxin Sun, Bin Shen, Jinsong Hu and Yang Yu

Abstract In rainy weather, traffic accidents are happening with increasing frequency. Due to the influence of weather and roads, driving vehicle often becomes dangerous, and it also affects the smoothness of the entire traffic system. After long years of study, it is found that the state of the drivers who drive in rainy weather is one of the important factors affecting the safety of the flow of traffic. In order to optimize the driving safety in rainy weather, based on the urban weather characteristics of the rainy season in Guangzhou, we test and analyze the physiological and psychological characteristics of drivers in rainy environment. The results show that the driving pressure and the mental load of drivers increase in rainy days, and the process of driving is accompanied by tension, panic, and insecurity by means of the comparison with the ECG index fluctuates between rainy weather and sunny weather. The research results will help the further optimization in the behavior of driving in rainy weather, so as to improve the smoothness and safety of traffic flow in rainy weather.

10.1 Introduction

In rainy weather, the visibility of the driver is reduced, and the road surface is slippery, which causes the difficulty in motor vehicle control. Driving in rainy weather has become one of the driving behaviors that cause frequent accidents. Therefore, the safety of driving in rainy weather has become one of the hottest issues that scholars at

W. Qi (✉) · Z. Sun · B. Shen
School of Civil Engineering and Transportation, South China University of Technology, Guangzhou 510641, China
e-mail: qwwhit@163.com

J. Hu
Guangzhou Transport Planning Research Institute, Guangzhou 510030, China

Y. Yu
School of Civil and Environmental Engineering, University of Technology Sydney, Ultimo, NSW 2007, Australia
e-mail: Yang.Yu-4@student.uts.edu.au

© Springer Nature Singapore Pte Ltd. 2019
X. Qu et al. (eds.), *Smart Transportation Systems 2019*, Smart Innovation, Systems and Technologies 149, https://doi.org/10.1007/978-981-13-8683-1_10

home and abroad have focused on. In the past study, sampling and modeling showed that the risk of driving on rainy days was much higher than which on sunny days. Besides, the probability of traffic accidents was two to three times than that in ordinary dry weather [3]. Taking Thailand as an example, due to the tropical monsoon climate, it is always in rainy season and there is a large amount of rainfall. The rainy season also leads to a lot of traffic accidents, and the accident rate is often higher than that in dry weather [12]. Another city, Sydney, which is the subtropical humid climate with frequent rainfall. Australia is heavily dependent on highways, especially from the Gold Coast to Brisbane, which relies on a large number of highway connections [14]. Studies showed that rainfall always implied driving risk, especially after the successive drought, as the drought continued, it would have a greater impact on the number of normal accidents [7]. From this point of view, ensuring the safety of rainy days has been a key issue for experts in domestic and international traffic for a long time.

The main reason for the danger of driving in rainy weather is that the road is slippery. When the speed is too high, the adhesion of the vehicle will decrease, it will cause the vehicle to slip and roll over. At the same time, the braking distance will be lengthened because of the excessive speed, thus the vehicle will hit another one and pedestrian in front. The visibility of pedestrians is also blurred in rainy days, which not only affects the safety of vehicles but also threatens the normal crossing of pedestrians [5]. Studies have shown that as a result of the formation of a thin layer of stagnant water on the road in rainy days, as a water film attached to the road surface, this "film" greatly reduces the friction coefficient between the tire and the road surface, and thus affects the road surface slip resistance [8]. In the statistics of traffic accidents in Nigeria, it is found that the traffic accidents caused by roads account for a large proportion through the establishment of the model. Because Nigeria is the tropical savanna climate, which is generally hot and rainy, so that the braking distance becomes longer in rainy weather [11]. Meanwhile, the power of braking is also greatly reduced in rainy weather. There are some experts who are analyzing the relationship between slip resistance, road characteristics, and collision based on various factors [13]. The rainy days have an impact on slippery roads, which makes the vehicle uncontrollable and difficult to predict.

The rainy weather will make vehicle travel affect the traffic flow as well, speed and density often determine the state of the entire traffic flow [19]. According to the survey, the traffic activities on weekdays during heavy rainfall will be less than 3%. However, transport operation on weekend will be reduced by more than 4% [6]. It will not only make influence on traffic generation, but also affect the efficiency of traffic, which causes traffic delays and shows a positive correlation [17]. At the same time, heavy rainfall will also lead to the reduction of traffic speed and maximum queuing flow at highway bottlenecks [16]. The main reason is the loss of visibility in rainy days, which leads to a decrease in speed and capacity, thus a slowdown in free-flow speed [4].

The traffic flow state of the expressway has certain adaptability to traffic interference. That is to say, in the harsh environment, the traffic flow state will have a certain resistance to traffic interference [9]. It is one of the participants in the traffic—the car, which is one of the most important factors affecting traffic flow, in other words, the driver. The traffic accident caused by the driver's subjective judgment and the wrong operation of the vehicle is the main component of the rainy traffic accident. It is more difficult for the driver who driving in rainy weather to find and identify the objects present in this environment than usual [15]. For example, the blurs of the windshield and rearview mirror, the light reflection of wet road, and so on. That is to say, the visibility of the driver is limited and the dynamic visual acuity is greatly reduced in rainy condition [1]. In addition, the attention of the drivers is also an important issue, and concentration can reduce the probability of traffic accidents [10]. At the same time, driving experience and driver's age are also the factors that affect driver behavior. According to research, high-density traffic flow and navigation challenges can lead to driver's mental stress and driving risk [18]. But experienced drivers are more comfortable than novices and can control the speed subconsciously. That is to say, the ability of driver to apperceive and handle decides whether it can be safe to drive under special weather conditions [2].

From the above, it seems that the driving situation in rainy weather is more complicated. And the external weather causes the abnormal operation of the traffic participants. For example, the vehicle has a weak adhesion because of the slippery of the road, thus a poorer braking performance and a longer braking distance. And because of the limitations of the driver's own physiological conditions in harsh environments, driver's psychological tension, and incompatibility of the environment. It causes a decline in the reaction power, and makes drivers overwhelmed. Moreover, traffic participants feel uncomfortable and drive more cautiously in rainy weather. It reduces the travel generation rate and the driving speed, thus congestion and limiting the traffic flow.

10.2 Analysis of Rainfall in Guangzhou

Guangzhou, located on the subtropical coast, is the maritime subtropical monsoon climate with a humid and rainy atmosphere. There are many rainy days throughout the year. From 2007 to the end of 2018, the average of annual rainy days in 12 years was about 152 days, accounting for more than 40% of the year. Since 2014, the annual rainy days have grown steadily, and the trend chart is shown in Fig. 10.1.

The precipitation in Guangzhou is abundant, and the flood season is generally from March to September. It is slightly different every year. The flood season begins in the time from March to May generally. According to the statistics of China Meteorological Administration, from 1981 to 2010, the average of yearly precipitation in Guangzhou was up to 1801.4 mm, and the average monthly precipitation in June was 318.6 mm.

The annual rainfall in Guangzhou is relatively concentrated. From the statistics of monthly precipitation in the past 12 years (as shown in Table 10.1). The annual rainfall was mainly concentrated from May to August (including the dragon boat water period). After September, it gradually decreased and tended to be flat. In addition,

Fig. 10.1 Statistics of rain days in Guangzhou from 2007 to 2018

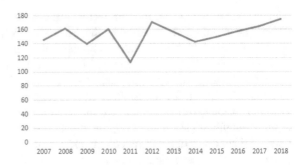

from January to February, Guangzhou was affected by cold air, and it continued to be in a transitional period from low temperature and rainy to sunny weather. The rainfall was generally stable. However, due to the influence of strong cold air in 2008, the transition period continued to be cold and rainy. The precipitation was one to three times as much as normal. Generally speaking, winter is a season of less rain in south China, with less precipitation probability and less rainfall. However, since 2013, there have been several rare winter rainstorms in Guangzhou, including December 2015, January 2016, and January 2018. In January 2016, rare winter rainstorms occurred one after another, and there is the first snowfall since the founding of the country in 2016. Even the monthly precipitation is up to 410.2 mm, which is much higher than the precipitation in the same period of the previous year.

At the same time, the rainfall in Guangzhou is closely related to the typhoon landing. From the data of the past 10 years (2008), typhoon weather came with a lot of precipitation and caused great interference to the road surface. In 2008, following the typhoon called "Neoguri" landing in April, "Fengshen" began to land in July, Guangzhou had a large range of heavy rains. The monthly precipitation is as high as 834.6 mm. And in 2010, there is only one typhoon "Chanthu" landed in Guangdong. But owing to the impact of the other five typhoons (including the strong tropical storm "Lionrock"), the annual precipitation is also as high as 2353.6 mm. On account of the strong typhoon "Nida" positive attack, Guangzhou released Red warning signal of the first typhoon in this century, the precipitation was 2939.7, that the highest since the meteorological began to record in 1908. In 2017, it was affected by six typhoons: it has seriously affected Guangzhou that "Hato", "Pakhar", and "Mawar" in half a month, bringing storms and strong storm surges. As a result of the uncertainty of the typhoon landing and impact, there are often sudden rainy seasons and attacks by heavy rainfall. Therefore, January, October, and December, non-rainy seasons, have become the seasons with a large amount of rainfall in recent years as well.

In virtue of frequent rainfall, range of visibility is limited in rainy weather, driving operation is closely related to the ability of dangerous sense and responsiveness. Intentional and negligent driving behavior is extremely dangerous. Therefore, it is very necessary to study the physiological and psychological characteristics of drivers in rainy weather. At the same time, rainy days in Guangzhou account for a large pro-

Table 10.1 Monthly precipitation in Guangzhou from 2007 to 2018

Year	2018	2017	2016	2015	2014	2013	2012	2011	2010	2009	2008	2007
1	133.4	13.4	410.2	55.9	0.7	3.8	72.6	26.6	69.5	5.4	98	23.9
2	17.1	26.4	41.8	40.9	33.7	8	70.8	45.2	71.7	0.6	49.9	25
3	64	174.1	253.8	27.2	274.6	174.2	63.9	49.3	19	207.7	70.9	44.5
4	101.3	117.7	272.6	116.4	177.1	282.8	365.2	27.9	256.7	108.8	111.7	184.9
5	172.3	421.4	297.5	805.6	542.9	300.6	291.1	185.3	584.5	210.9	285.2	193.7
6	574.8	405.8	520.1	251.8	300.7	228.2	302.9	493.8	322.7	273.7	834.6	319.8
7	193.4	312	301.2	441.2	199.8	318.9	136.6	363.5	164.9	221.6	170.3	80.7
8	318.8	244.8	425.6	342.7	513.1	395.5	184	37.4	250.5	230.1	188.4	309.5
9	216.8	269.4	210.8	116.4	96.1	231	69.4	183.9	558.6	39	262.6	151.5
10	82.5	44.3	140.9	123	1.2	5	42.8	154.7	28.9	26.4	136.4	18.6
11	33.1	36.7	61.8	44.7	35.9	42.2	168.8	64.6	2.1	101.2	61.9	2
12	18.7	1.4	3.4	106.1	58.2	105.2	45.8	0.1	24.5	47.2	14.1	16.2
Total	1926.2	2067.4	2939.7	2471.9	2234	2095.4	1813.9	1632.3	2353.6	1472.6	2284	1370.3

portion, rainfall intensity improves and rainy season is very frequent and changeable, which is very conducive for us to conduct this research based on this.

10.3   Experimental Design and Data Reduction

10.3.1   Description of the Simulated Driving Experiment

This experiment adopts the experimental method of simulated driving to compare and study the change rules of physiological and psychological characteristics of drivers in the driving process in rainy and sunny conditions. The experimental equipment mainly includes automobile driving simulator for road safety (as shown in Fig. 10.2) and MP36R four-channel physiological instrument. It can simulate the road traffic and environmental scene through the driving simulator, and the traffic flow is free flow. The driver can obtain the same visual effects and driving experience as in the actual road driving in the driving device. The MP36R four-channel physiology instrument can record the driver's data of ECG, EEG, and respiration during driving.

10.3.2   Participant and Procedure

In this experiment, 10 drivers were selected as test subjects. All the drivers involved in the test held motor vehicle driving licenses and already had rich driving experience in various types of roads. One hour was given for the operation of the driving simulator to get them familiar with it before the formal test to eliminate the strangeness of simulation driving platform and ensure the reliability of the experimental data. Before

Fig. 10.2   Automobile driving simulator for road safety

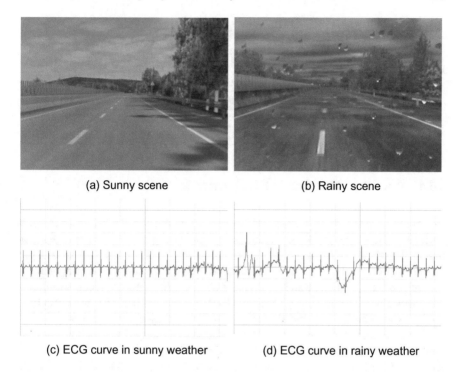

(a) Sunny scene   (b) Rainy scene

(c) ECG curve in sunny weather   (d) ECG curve in rainy weather

Fig. 10.3 Sunny and rainy scene and associated data characterization

the test, the driver was familiar with various simulation scenarios and had a complete driving experience.

The test section of this experiment was an expressway section with a central isolation zone, no non-motor vehicles and pedestrians, and twin four-lane. The test environment was a normal traffic condition, that is to say a free-flow driving state, and there were never any other vehicles to interfere with the normal operation of the test vehicle. The rainfall environment was the system's default environment with moderate rain.

Before the start of the experiment, the recorder strictly checked the experimental equipment to prevent other accidental factors from interfering with the experiment. The recorder synchronously recorded the time of the driving simulation system switching scene and the moment of the multi-channel physiological instrument, and observed the input stability of physiological index data in notebook every 1 min. The visualization of the initial data is shown in Fig. 10.3.

Table 10.2 Time-domain Index of driver heart rate variability

Time-domain index	Normal		Rainy days	
	Mean value	Standard deviation	Mean value	Standard deviation
MEAN	0.649328	0.075926	0.696231	0.070083
SDNN	0.056104	0.030469	0.081858	0.067626
RMSSD	0.065652	0.060379	0.111037	0.115156
SDSD	0.065501	0.060424	0.111025	0.115155

10.4 Analysis of Driver's ECG Index in Rainy Weather

The analysis of the driver's ECG index is mainly to analyze the driver's heart rate variability. Heart rate variability refers to the small changes in the heart rate fluctuation cycle, which usually refers to the smile fluctuations of the RR interval between successive heartbeats. Heart rate variability reflects the tension and balance of activities of cardiac sympathetic and parasympathetic, and has a good effect in evaluating the driver's autonomic activity. The analysis of heart rate variability is mainly carried out by two methods: time-domain analysis and frequency-domain analysis.

The time-domain analysis of the driver's heart rate variability is mainly based on various statistical analysis methods to quantitatively describe the variation characteristics of the RR interval, usually characterized by MEAN, SDNN, RMSSD, and SDSD. Driver's time-domain indexes after the data processing were sorted out to Table 10.2.

According to the data in Table 10.2, the MEAN, SDNN, RMSSD, and SDSD of the driver in rainy weather increased greatly compared with the normal driving environment, and it shows that the change of RR interval value becomes larger, sympathetic activity and the mental stress of drivers increase. That is to say, in rainy environment, the driver is unclear in vision, the road surface is slippery, the pressure of driver increases, and it is easy to feel fatigue for driver. With the increase of the possibility of misjudgment and operation greatly, it is dangerous to drive with this condition.

The frequency-domain analysis, which is same as power spectral analysis, is to decompose the change signal of heart rate into different frequencies by appropriate arithmetical operation. It is related to time-domain analysis, but it can explain more complex changes. This paper mainly uses the LF/HF of HRV to judge the mental state of the driver in rainy environment. Driver's frequency-domain indexes after the data processing were sorted out to Table 10.3.

In the normal driving environment, the low-frequency and high-frequency ratio (LF/HF) of the driver HRV is reasonable. This indicates that the driver's mental condition is relatively stable, driving stability is enhanced, and driver is less prone to accidents. In rainy environment, the LF/HF of the HRV is increased. This indicates that the driver's mental condition is unstable in this condition, the mood is easy to up and down, the driving pressure increases as well, and the driving stability is

Table 10.3 Frequency-domain index of driver heart rate variability

Frequency-domain index	Normal		Rainy days	
	Mean value	Standard deviation	Mean value	Standard deviation
LF	0.201882	0.039327	0.226812	0.09143
HF	0.798118	0.039327	0.773188	0.09143
LF/HF	0.255924	0.060495	0.312759	0.165059

lowered. On the one hand, in rainy environment, the sight line orientation of the driver is blocked, the driver is difficult to fully grasp the situation ahead, leading to the increase of driving pressure. On the other hand, both road conditions and vehicle conditions are reduced in rainy weather, making it more difficult for drivers to drive in rainy weather, leading to nervousness of drivers and decreasing driving stability.

10.5   Summary

According to the above research, the driving pressure increases in rainy environment, and the mental load is improved. In the process of driving, there are senses of tension, panic, and insecurity, and the ECG index fluctuates compared with that in the sunny environment. That is to say, during driving in rainy weather, the driver's operational stability is degraded because of the changes of characteristics in vehicle, road, and traffic flow; besides, the driver's control ability of the surrounding is reduced. The driver is in a kind of unstable state is prone to fatigue and has the possibility of causing traffic accidents. To this end, on the one hand, drivers need to reduce this risk by slowing down, increasing the mental effort involved in driving tasks, driving more carefully, maintaining stable driving, and so on. On the other hand, we also need to take some traffic control measures to guide drivers to complete driving tasks safely, such as speed limit, ban, etc.

Acknowledgements The study is supported by the National Natural Science Foundation of China (No. 71701070 & 71701041), the Natural Science Foundation of Guangdong Province (No. 2016A030310427), and the Science and Technology Project of Guangzhou City (No. 201804010466).

References

1. Ahmed, M.M., Ghasemzadeh, A.: The impacts of heavy rain on speed and headway behaviors: an investigation using the SHRP2 naturalistic driving study data. Transp. Res. Part C Emerg. Technol. **91**, 371–384 (2018)
2. Bo, L., Yingzi, L.: Using physiological and behavioral measurements in a picture-based road hazard perception experiment to classify risky and safe drivers. Transp. Res. Part F Traffic

Psychol. Behav. **58**, 93–105 (2018)

3. Brodsky, H., Hakkert, A.S.: Risk of a road accident in rainy weather. Accid. Anal. Prev. **20**(3), 161–176 (1988)
4. Camacho, F.J., García, A., Belda, E.: Analysis of impact of adverse weather on freeway free-flow speed in Spain (2010)
5. Easa, S.M., Qu, X., Dabbour, E.: Improved pedestrian sight distance needs at railroad-highway grade crossings. J. Transp. Eng. Part A Syst. **143**(7), 04017027 (2017)
6. Hassan, Y.A., Barker, D.J.: The impact of unseasonable or extreme weather on traffic activity within lothian region, Scotland. J. Transp. Geogr. **7**(3), 0–213 (1999)
7. Keay, K., Simmonds, I.: The association of rainfall and other weather variables with road traffic volume in Melbourne, Australia. Accid. Anal. Prev. **37**(1), 109–124 (2005)
8. Kokkalis, A.G., Panagouli, O.K.: Fractal evaluation of pavement skid resistance variations. i: surface wetting. Chaos Solitons Fractals **9**(11), 1875–1890 (1998)
9. Kuang, Y., Qu, X., Wang, S.: A tree-structured crash surrogate measure for freeways. Accid. Anal. Prev. **77**, 137–148 (2015)
10. Narad, M., Garner, A.A., Brassell, A.A., Saxby, D., Antonini, T.N., O'Brien, K.M., et al.: Impact of distraction on the driving performance of adolescents with and without attention-deficit/hyperactivity disorder. JAMA Pediatr. **167**(10), 933 (2013)
11. Osayomi, T.: Regional determinants of road traffic accidents in Nigeria: identifying risk areas in need of intervention. Afr. Geogr. Rev. **32**(1), 88–99 (2013)
12. Pitaksringkarn, J., Tanwanichkul, L., Yamthale, K., Hamontree, C.: A correlation between pavement skid resistance and wet-pavement related accidents in Thailand. MATEC Web of Conferences, p. 192 (2018)
13. Piyatrapoomi, N., Weligamage, J., Kumar, A., Bunker, J. M.: Identifying relationship between skid resistance, road characteristics and crashes using probability-based risk approach (2008)
14. Qu, X., Wang, S.: Long distance commuter lane: a new concept for freeway traffic management. Comput. Aided Civ. Infrastruct. Eng. **30**(10), 815–823 (2015)
15. Sato, R., Domany, K., Deguchi, D., Mekada, Y., Ide, I., Murase, H., et al.: Visibility estimation of traffic signals under rainy weather conditions for smart driving support. International IEEE Conference on Intelligent Transportation Systems. IEEE (2012)
16. Seeherman, J.L.: The impact of adverse weather on freeway bottleneck performance. Dissertations & Theses—Gradworks (2014)
17. Tsapakis, I., Cheng, T., Bolbol, A.: Impact of weather conditions on macroscopic urban travel times. J. Transp. Geogr. **28**, 204–211 (2013)
18. Trick, L.M., Toxopeus, R., Wilson, D.: The effects of visibility conditions, traffic density, and navigational challenge on speed compensation and driving performance in older adults. Accid. Anal. Prev. **42**(6), 1661–1671 (2010)
19. Zhang, J., Qu, X., Wang, S.: Reproducible generation of experimental data sample for calibrating traffic flow fundamental diagram. Transp. Res. Part A **111**, 41–52 (2018)

Chapter 11
Identification of Factors Influencing Crash Severity for Electric Bicycle Using Nondominated Sorting Genetic Algorithm

Cheng Xu

Abstract Electric bicycles (E-bike) are one of the most important travel modes in China. In recent years, traffic accidents involving electric bicycles have increased year by year, and research on traffic safety risks of electric bicycles is particularly important. The key factor in obtaining traffic accidents involving electric bicycles is an important basis for the development of electric bicycle traffic management and the relevant policies. Therefore, based on the electric bicycle traffic accident in Hangzhou, this paper uses the nondominated sorting genetic algorithm II (NSGA-II) to study the key factors affecting the severity of electric bicycle accidents. The results show that the type of accident and the type of illegality are the two most important factors affecting the severity of electric bicycle accidents.

11.1 Introduction

With the increase in travel distance in many Chinese cities, due to the advantages of low cost, flexible operation, and easy riding, E-bikes have become an important way of travel. On the one hand, electric bicycles have high efficiency and are an important green way of travel. On the other hand, accidents involving electric bicycles have also increased dramatically. In Zhejiang Province, for example, in 2017 there were 1146 fatalities involving electric bikes. An increase of 3.15% over the same period of last year, accounting for 28.8% of the total number of traffic crashes. Electric bicycles have become a "pain point" for traffic safety. Therefore, it is important to identify factors influencing crash severity for electric bicycles and provide strong data support for E-bikes traffic management and crash prevention.

C. Xu (✉)
Department of Traffic Management Engineering, Zhejiang Police College, Hangzhou 310053, China
e-mail: xucheng@zjjcxy.cn

Institute of Intelligent Transportation Systems, Zhejiang University, Hangzhou 310058, China

© Springer Nature Singapore Pte Ltd. 2019
X. Qu et al. (eds.), *Smart Transportation Systems 2019*, Smart Innovation, Systems and Technologies 149, https://doi.org/10.1007/978-981-13-8683-1_11

In recent years, research on safety risks and traffic flow theory of E-bikes have been rapidly developed [1–5]. Hu et al. [6] analyzed the severity of the accident and its influencing factors through the data of electric bicycle traffic accidents in Hefei, China. Kim et al. [7] analyzed bicycle and motor vehicle collision accidents and their severity. Weter et al. [8] evaluated the severity and influencing factors of electric bicycle accidents in Switzerland through actual case analysis. Bambach et al. [9] analyzed the impact of drivers' safety helmets on bicycle traffic safety. Langford et al. [10] based on GPS technology analysis of hybrid bicycle safety behavior in four cases and found that electric bicycles and bicycles do not drive on a regular basis and violations of signal signs marking traffic violations of these two types of violations rates are high. Based on the questionnaire survey of e-bike users distributed in Denmark, Haustein and Moller [11] concluded that the riding style and attitude of E-bike drivers play a crucial role in major traffic crashes.

From the references, we can see that foreign research lacks a large number of data on electric bicycle traffic accidents. In this paper, we take the data of traffic crashes involving electric bicycles occurring in Xintang Street, Hangzhou, Zhejiang Province from June 2015 to May 2016 as a sample, and use NSGA-II to identify the key factors affecting the severity of electric bicycle crashes. According to the result of NSGA-II, this paper puts forward the countermeasures for the traffic management of electric bicycles.

11.2   Crash Data of Electric Bicycle

11.2.1   Crash Data Source

Since a large number of crashes involving electric bicycles are handled by simple procedures, it is often the case that many crashes caused by electric bicycles are missed by analyzing only general procedures for dealing with crashes. Therefore, the text adopts the summary procedure to deal with crash and general procedure to deal with crash data from 0:00:20, June 1, 2015 to 24:00, May 31, 2016 in the area of Xintang Subdistrict, Hangzhou.

The crash database category includes the crash occurrence time, weather, location, cause of the incident, personal injury and property damage, degree of loss, crash liability, crash type, age of electric bicycle driver, gender, and other related data. Through the above analysis of various factors, we can conduct in-depth analysis of the causes of traffic crashes involving electric bicycles and provide a reference for the prevention of traffic crashes.

11.2.2 Characteristics of Crashes

E-bike traffic crash analysis area is in Xintang streets. The total area of 34.8 km^2, resident population of 65,000, more than 140,000 nonlocal population. As the area is located at the junction of urban and rural areas, there are a large number of migrant workers, high E-bike ownership in the area, and heavy demand for commuting during peak periods. The traffic conflicts between electric bicycles and other vehicles are serious and the traffic safety situation is severe.

From the time point of view, electric bicycle traffic crashes show the phenomenon of time is not balanced. The time distribution of traffic crashes in different months and different time periods, respectively. In the distribution of the month, the number of electric bicycle crashes in a year except for January–February due to the Spring Festival and other reasons there is a clear downward trend, as well as a marked increase in March, the number of traffic crashes in other months was basically stable. In the period distribution, the distribution of electric bicycle crashes showed obvious saddle type, basically the same with the time distribution characteristics of traffic demand, and the rapid increase of traffic brought the rapid growth of traffic crashes.

It can be seen from the data that the highest percentage of cyclists in the 40–50 age group should be closely related to the number of travelers in this group. In terms of gender, male cyclists accounted for 58.39% of all incidents and summary cyclists handled 41.61% of female cyclists. In terms of types of crashes, the number of different types of crashes, with motor vehicles and electric bicycle crashes and electric bicycle crashes accounting for 79.84 and 17.42%, accounting for the absolute majority. These two types of crashes are also the main causes of injuries to electric bicyclists.

11.3 Identification of Factors Influencing Crash Severity

11.3.1 Definition of Values

The data from the accident dataset is prepared, and nine explanatory variables were considered for the analysis. Time, cyclist, environment, and crash are the four categories of those variables. Table 11.1 shows the grouping and values definition for all the nine candidate variables considered. The factor Time consists of three variable such as Day of Week (F-1) with 2 values (Weekday and Weekend), Time (F-2) with 24 values, and Month (F-3) with 12 values. The factor Cyclist consists of two variables such as Cyclist Age (F-4) with three values Young, Middle-aged, and Elderly; Cyclist Gender (F-5) with two values, Male and Female. The factor Environment consists

Table 11.1 Definition of values for all the candidate factors from four categories

Category	Factor num.	Candidate factor	Factor values definition
Time	F-1	Day of week	(1) Weekday. (2) Weekend
	F-2	Time	(1) 0–1 o'clock (2) 1–2 o'clock ……
	F-3	Month	(1) Jan. (2) Feb. ……
Cyclist	F-4	Cyclist age	(1) Young. (2) Middle-aged. (3) Elderly
	F-5	Cyclist gender	(1) Male. (2) Female
Environment	F-6	Weather	(1) Sunny day (2) Cloudy day (3) Rain and snow
	F-7	Road form	(1) Intersection (2) Road section
Crash	F-8	Crash type	(1) Electric bicycles and electric bicycles (2) Electric bicycles and motor vehicles (3) Electric bicycles and bicycles (4) Electric bicycles and pedestrian (5) Electric bicycle unilateral crash
	F-9	Illegal behavior	(1) Motor vehicle illegal (2) Non-motor vehicle driving in reverse (3) Driving electric bicycles in violation of the provisions of manned (4) Non-motor vehicles are not driving in the non-motor vehicle lane (5) Non-motor vehicles do not drive on the right side of the road (6) Turn nonmotorized vehicles to keep going straight (7) Non-motor vehicles violate traffic signal regulations (8) Other non-motor vehicle offenses

of two variables such as Weather (F-6) with 3 values (Sunny day, Cloudy day, and Rain and snow); Road form (F-7) with 2 values (Intersection and Road section). The factor Crash consists of two variables such as Crash type (F-8) with 5 values (Electric bicycles and electric bicycles, Electric bicycles and motor vehicles, Electric bicycles and bicycles, Electric bicycles and pedestrian, and Electric bicycle unilateral crash); Illegal behavior (F-9) with 8 values (Motor vehicle illegal, Non-motor vehicle driving in reverse, Driving electric bicycles in violation of the provisions of manned, Non-motor vehicles are not driving in the non-motor vehicle lane, Non-motor vehicles do not drive on the right side of the road, Turn nonmotorized vehicles to keep going straight, Non-motor vehicles violate traffic signal regulations and other non-motor vehicle offenses).

The severity of traffic crashes is one of the main indicators of the severity of the crash. Therefore, in-depth analysis of the severity of the crash is helpful to identify the main factors causing the crash injury and provide the data support for further reducing the crash injury. In order to further analyze which factors will affect the severity of electric bicycle traffic crashes, we classify the statistics of the rider's gender, age, crash time, month, weather, road form, crash type, and illegal behavior as shown in Table 11.1.

11.3.2 Optimization Model

For the optimization model, nondominated sorting genetic algorithm II (NSGA-II) is proposed to efficiently search for optimal solutions.

First, in the optimization model, the set of all the optimal solutions known as the Pareto set is found by NSGA-II algorithm. This set is used to determine the optimal number of significant factors. Next, a quantifying index is defined from the concept of Schema in a genetic algorithm, and the corresponding factor significance (Fs) values are calculated based on the important gene analysis technique of the Pareto set. Then, the performance of the selected factors is used to create a basic pattern and defines the most important significant factors.

This research proposed NSGA-II model for the optimization of factors influencing crash severity by selecting fewer factors (g_1) with higher identifying accuracy (g_2), with g_1 and g_2 as the two optimization objectives. The main procedure of a MOP problem can be described as an iterative loop between a decision space X and an objective space Y. A function f transfers the decision space into the objective space as $f: X \rightarrow Y$. Next, evaluation and search of the results in the objective space are performed to provide feedback to the decision space.

To formulate the identification of significant factors in a MOP standpoint, the decision-vector is set as $\vec{x} = (x_1, x_2, \ldots, x_9)$, where

$$x_i = \begin{cases} 1, & i\text{th variable selected;} \\ 0, & i\text{th variable unselected;} \end{cases} \quad i = 1, 2, \ldots, 9. \quad (11.1)$$

The first objective function is $f_1(x) = g(D(x))$, where D(x) is the MSE of the neural network output from a dataset with given factor combination. The second objective function is the total number of the selected variables as $f_2(x) = \sum_{i=1}^{9} x_i$. Thus, the MOP problem in this research can be formulated as follows:

$$\min_{x \in X} f(x) = (f_1(x), f_2(x))^T \quad (11.2)$$

$$s.t. f_1(x) = g(D(x)) \quad (11.3)$$

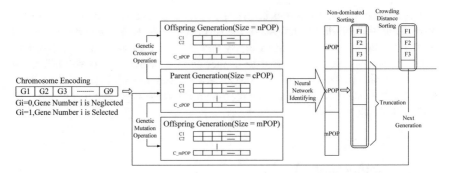

Fig. 11.1 Description of the main steps of the NSGA-II model implementation

$$f_2(x) = \sum_{i=1}^{9} x_i \tag{11.4}$$

$$X = \left\{ x \in R^n / x_i(x_i - 1) = 0, \ i = 1, 2, \ldots, n \right\} \tag{11.5}$$

In this research, we employed NSGA-II, a generic multi-objective algorithm used to search for nondominated solutions. This model has better convergence properties because it stores all nondominated solutions. It also adapts a suitable automatic mechanism based on the crowding distance to guarantee the diversity and spread of its solutions. Figure 11.1 shows the main steps in the implementation of the model. To implement the model, the chromosome of the NSGA-II is encoded as a binary sequence of 9 bits. Every bit stands for a corresponding factor and the whole bit-string represents a certain combination of the candidate phone-use factors. For the initialization of the population, a parent generation that includes a size of nPOP chromosomes is randomly generated. In each loop of the following iterations, two offspring populations are generated from corresponding parent generation through genetic operators of crossover and mutation. The size of these two offspring populations is cPOP and mPOP, respectively. For each individual in the simulation population, all the "1" bit fields in the chromosome will be retrieved from the original data set and connected to the NN input. All the crash data sample is divided into training set, validation set and test set with the proportion ratio of 70, 15, and 15%, which can prevent the model from over fit during the training period. Then based on the output of the two objective functions, better-qualified chromosomes are chosen according to the NDS and CD through the NSGA-II algorithm. At the end of the simulation of the iterations, the model converges to the best chromosomes that represent the optimal or suboptimal solutions. NSGA-II can converge to the optimal solution or the approximate optimal solutions very fast.

Table 11.2 Pareto optimal solution set

Number of variable	1	2	3	4	5	Last results
1	0001000000	0001000000	0001000000	0001000000	0001000000	0001000000
2	0001001000	0001001000	0001001000	0001001000	0001001000	0001001000
3	0001001001	0001101000	0011001000	0001001010	0001001001	0001001001
4	1001001001		1001000101	0011001010	1001001001	1001001001
5	1111000001		0011011100	1101101000	1011001001	
6	1001101101	0011001111	0111101001	1011011010	0111001101	
7	1011111001		1111111000	0111011101	1111001101	
8	1011111101		1111001111	0111011111		
9	1011111111	1111111011	1111011111			
10	1111111111		1111111111	1111111111		

11.3.3 Results

Change the numerical change of the output class to one hot code and run NSGA-II five times. Record the Pareto optimal solution set after convergence of each iteration and the results can be shown in Table 11.2.

For illustration, Fig. 11.2 shows the change in the number of nondominant fronts as the NSGA-II iteration goes on by performing the simulations with the data set in this study. It shows that the simulation was able to converge to the optimal solution at the fourteenth iteration. When the number of parameters is 2, the minimum error is achieved. After finding the optimal combination, with the number of parameters increases, and there is no obvious rule for the selected optimal combination. Therefore, the most two critical factors influencing crash severity of electric bicycles are crash type and illegal behavior.

11.4 Conclusions

The goal of this paper is to give information about the determinants, or rather the most important ones that influence the crash severity of electric bicycles. We proposed a NSGA-II model that adequately fit well the data, to identify the most important significant covariates with prevalence evaluation. Due to the strong correlation between factors, data in this study reveal that crash type and illegal behavior are the top two significant and important factors in crash severity of electric bicycles. The research suggests to take traffic safety management measures to further reduce the illegal activities of electric bicycles.

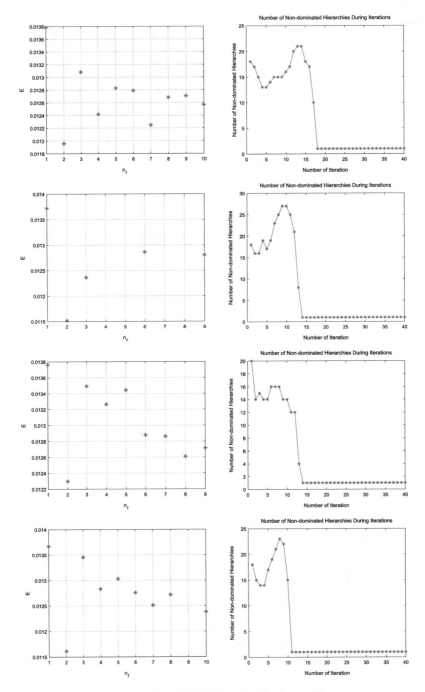

Fig. 11.2 Convergence process of the NSGA-II (Iteration Number = 40)

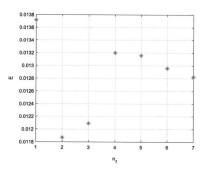 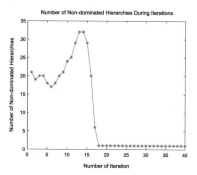

Fig. 11.2 (continued)

Acknowledgements This work was supported by the Zhejiang Provincial Natural Science Foundation of China (LQ17E080001), and the China Postdoctoral Science Foundation.

References

1. Jin, S., Qu, X., Zhou, D., Xu, C., Ma, D., Wang, D.: Estimating cycleway capacity and bicycle equivalent units for electric bicycles. Transp. Res. Part A **77**, 225–248 (2015)
2. Wang, D., Ma, X., Ma, D., Jin, S.: A novel speed-density relationship model based on the energy conservation concept. IEEE Trans. Intell. Transp. Syst. **18**, 1179–1189 (2017)
3. Jin, S., Luo, X., Ma, D.: Determining the breakpoints of fundamental diagrams. IEEE Intell. Transp. Syst. Mag. **11**, 110–120 (2019)
4. Qu, X., Zhang, J., Wang, S.: On the stochastic fundamental diagram for freeway traffic: model development, analytical properties, validation, and extensive applications. Transp. Res. Part B **104**, 256–271 (2017)
5. Xu, C., Yang, Y., Jin, S., Qu, Z., Hou, L.: Potential risk and its influencing factors for separated bicycle paths. Accid. Anal. Prev. **87**, 59–67 (2016)
6. Hu, F., Lv, D., Zhu, J., Fang, J.: Related risk factors for injury severity of e-bike and bicycle crashes in Hefei. Traffic Inj. Prev. **15**, 319–323 (2014)
7. Kim, J.K., Kim, S., Ulfarsson, G.F.: Bicyclist injury severities in bicycle-motor vehicle accidents. Accid. Anal. Prev. **39**, 238–251 (2007)
8. Weber, T., Scaramuzza, G., Schmitt, K.-U.: Evaluation of e-bike accidents in Switzerland. Accid. Anal. Prev. **73**, 47–52 (2014)
9. Bambach, M.R., et al.: The effectiveness of helmets in bicycle collisions with motor vehicles: A case-control study. Accid. Anal. Prev. **53**, 78–88 (2013)
10. Nordback, K., Marshall, W.E., Janson, B.N.: Bicyclist safety performance functions for a U.S. city. Accid. Anal. Prev. **65**, pp. 114–122 (2014).
11. Carter, D.L., Hunter, W.W., Zegeer, C.V., Stewart, J.R., Huang, H.: Bicyclist intersection safety index. Transp. Res. Rec.: J. Transp. Res. Board **2031**, 18–24 (2007)

Chapter 12
Traffic Signal Timing via Parallel Reinforcement Learning

Qian Zhao, Cheng Xu and Sheng jin

Abstract Nowadays, reinforcement learning is widely used to design intelligent control algorithms, which has gradually become one of the popular methods of signal control. We propose a new traffic signal control method, which applies parallel reinforcement learning methods to build a traffic signal control agent in the traffic micro-simulator. This method uses covariance adaptive matrix evolution strategy (CMA-ES) algorithm to train our system on computer cluster, with over 300 iterations and 500 populations in each iteration. In this paper, we provide preliminary results on how the parallel reinforcement learning methods perform in traffic signal control system.

12.1 Introduction

Many cities are suffering from the congestion traffic problem, which contains complexity and variability in various scenes and remains a hard problem for researchers. Many problems are caused by congestion, such as more wasting for petrol, more waiting time, more pollution and more accidents. Usually, researchers have tried to solve the congestion traffic problem by widening lanes, but it is impossible in some cities, which is too expensive. One of the best solutions is to optimize traffic signals and traffic lights. In many cities, traffic signals are controlled by camera or sensor agent which opposite state on demand or change the signals form red to green, however, not all cities have the ability to assemble these devices, in which case, pretimed method is widely used in traffics lights. It is necessary for us to select the optimal time for each traffic intersection, as a result, the congestion problem may be reduced.

Q. Zhao · C. Xu (✉) · S. jin
Institute of Intelligent Transportation Systems, Zhejiang University, Hangzhou 310058, China
e-mail: xucheng@zjjcxy.cn

Q. Zhao
e-mail: qianzhaoapollo@163.com

C. Xu
Department of Traffic Management Engineering, Zhejiang Police College, Hangzhou 310053, China

© Springer Nature Singapore Pte Ltd. 2019
X. Qu et al. (eds.), *Smart Transportation Systems 2019*, Smart Innovation, Systems and Technologies 149, https://doi.org/10.1007/978-981-13-8683-1_12

However, it is a very complex to optimize traffic light signal, in which the major two difficulties are the modelling difficulty and optimization difficulty. In this paper, we propose a new traffic signal control method, which applies parallel reinforcement learning methods to build a traffic signal control agent in the traffic micro-simulator, which efficiently manages the traffic according to its condition.

RL has been widely applied for adaptive traffic signal control, and for the comprehensive survey of existing methods in traffic can be referred to [1, 2]. Some researchers employed model-based RL in traffic signal control [3, 4]. Model-free RL methods in designing adaptive traffic signal control were also used [5–7]. Besides, multi-agent RL architectures were also proposed [8, 9] to address traffic signal control problems.

It is well-known that traffic signal timing problem is a multidimensional optimization problem, in which parallel reinforcement learning methods have not been comprehensively studied. In this paper, we analyze these issues in-depth and propose a simulation method based on the parallel reinforcement learning technology.

12.2 Algorithm of the CMA-ES

Covariant Matrix Adaptation with Evolution Strategy (CMA-ES) [10] is an improved robust form of minimization strategy, in which the main feature is the ability that it is invariant to landscape transformations and scaling modulation. Besides, it offers no discrepancy in behaviour towards varied nature of functions and is easily generalizable [11]. With updating schemes of CMA-ES, complexity of algorithm can be largely reduced [12, 13].

First, in CMA-ES, the offspring of 'λ' is updated by virtue of Eq. (12.4):

$$x_k^{(g+1)} \sim N(x_\mu^g, \sigma^{(g)^2} C^{(g)}), k = 1, \dots, \lambda \tag{12.1}$$

where

$$\langle x \rangle_\mu^{(g)} = \frac{1}{\mu} \sum_{i \in I_{sel}^{(g)}} x_i^{(g)} \tag{12.2}$$

Represents the weighted mean of the selected individuals of generation g, and $I_{sel}^{(g)}$ is the set of the indices, $N(\mu, C)$ represents a normally distributed random vector with mean μ and covariance matrix C.

The update Eq. (12.6) is

$$N(x_\mu^g, \sigma^{(g)^2} C^{(g)}) \sim x_\mu^{(g)} + \sigma^{(g)} B^{(g)} D^{(g)} N(0, I) \tag{12.3}$$

$B^{(g)} D^{(g)}$ represents random vectors, and the covariance matrix $C^{(g)}$ of $B^{(g)} D^{(g)}$ is a symmetrical positive $n \times n$ matrix.

Second, the covariance matrix $C^{(g)}$ is adapted by the evolution path $p_c^{(g+1)}$, and $x_W^{(g)}$ denotes the recombination point, and as seen in Eq. (12.1) which denotes the separation of present parents with $x_W^{(g)}$.

$$p_c^{(g+1)} = (1 - c_c) p_c^{(g)} + H_\sigma^{(g+1)} \sqrt{c_c(2 - c_c)} \frac{\sqrt{\mu_{eff}}}{\sigma^{(g)}} (x_W^{(g+1)} - x_W^{(g)}) \qquad (12.4)$$

$$C^{(g+1)} = (1 - c_{cov}) C^{(g)} c_{cov} \frac{1}{\mu_{cov}} p_c^{(g+1)} (p_c^{(g+1)})^T$$

$$+ c_{cov} \left(1 - \frac{1}{\mu_{cov}}\right) \sum_{i=1}^{\mu} \frac{w_i}{\sigma^{(g)}} (x_{i:\lambda}^{(g+1)} - x_{i:\lambda}^{(g)}) (x_{i:\lambda}^{(g+1)} - x_{i:\lambda}^{(g)})^T \qquad (12.5)$$

where

$$H_\sigma^{(g+1)} = 1, \text{ if } \frac{p_\sigma^{(g+1)}}{1 - (1 - c_\sigma)^{2(g+1)}} < \left(1.5 + \frac{1}{n - 0.5}\right) E(||N(0, 1)||) = 0, \text{ other-}$$

wise $\mu_{eff} = \frac{\left(\sum_{i=1}^{\mu} w_i\right)^2}{\sum_{i=1}^{\mu} w_i^2}$ represents "variance effective selection mass". $c_{cov} \approx \min\left(1, \frac{2\mu_{eff}}{n^2}\right)$ represents the covariance matrix C learning rate.

Third, to adapt global step size σ, the evolution path $p_\sigma^{(g+1)}$ is computed similar to the evolution path $p^{(g+1)}$ with the difference that $p_\sigma^{(g+1)}$ is not scaled by $D^{(g)}$ [14].

$$p_\sigma^{(g+1)} = (1 - c_\sigma) p_\sigma^{(g)} + \sqrt{c_\sigma(2 - c_\sigma)} B^{(g)} D^{(g)-1} B^{(g)T} \times \frac{\mu_{eff}}{\sigma^{(g)}} (x_W^{(g+1)} - x_W^{(g)}) \qquad (12.6)$$

The step size for generation $g + 1$ is determined by the length of the evolution path [15].

$$\sigma^{(g+1)} = \sigma^{(g)} \cdot \exp\left(\frac{1}{d_\sigma} \frac{||p_\sigma^{(g+1)}|| - \hat{\chi}_n}{\hat{\chi}_n}\right) \qquad (12.7)$$

where $\hat{\chi}_n = E[||N(0, I)||]$, which is the expected length of a $N(0, I)$ vector and d_σ is the damping parameter which is greater than one. And $\hat{\chi}_n$ is approximated by $\hat{\chi}_n \approx \sqrt{n}\left(1 - \frac{1}{4n} + \frac{1}{21n^2}\right)$ [14].

Parameter setting of the strategy has been fully discussed [14], and the default setting is shown as follows:

$$c_c = \frac{4}{n + 4}, c_{cov} = \frac{2}{(n + \sqrt{2})^2}, c_\sigma = \frac{4}{n + 4}, d_\sigma = c_\sigma^{-1} + 1 \qquad (12.8)$$

The initial covariance matrix $C^{(0)} = I$, and the evolution path for σ: $p_c^{(0)} = 0$, the evolution path for C: $p_\sigma^{(0)} = 0$.

12.3   System Description

In this section, we formulate traffic light control problem as a parallel reinforcement learning task by describing optimization mechanism as shown in Fig. 12.2. Then present how to train the system.

12.3.1   Environment and Input Traffic Parameters

The experiment is established in the micro-simulator. The specific input traffic parameters can be found in Table 12.1. Total phase state indicates the traffic signal state, the duration represents how long has the light been in each signal, and left turn bay occupation denotes the possibility of spillback. The reason why these parameters were selected is that these parameters can change the traffic state directly, which can be seen as an action set.

12.3.2   Fitness Function

The fitness value is a scalar value which received after each simulation cycle, each simulation cycle is active, one at a time for a fixed period of time, and the agent is rewarded based on fitness function. The following equation specifies the rewards, t_i is the travel time of vehicle, $t_{average}$ and $l_{average}$ is the average travel time and queue length of each vehicle, N denotes the total number of vehicles in each simulation period.

$$t_{average} = \frac{1}{N} \sum_{i=0}^{N} t_i, i = 1, 2, \ldots, N \tag{12.9}$$

$$l_{average} = \frac{1}{N} \sum_{i=0}^{N} l_i, i = 1, 2, \ldots, N \tag{12.10}$$

Table 12.1 Specific input traffic parameters

Notation	Description
Phase state	Total phase state
Average queue length of each lane	The average length that vehicles waiting at the intersection
Average cumulative delay	The average cumulative delay that vehicles waiting at the intersection
Green duration	How long has the light been in green, for each signal head

$$Fitness = \alpha * t_{average} + \beta * l_{average} \tag{12.11}$$

As explained in Eq. 12.11, fitness function is the liner combination of average travel time and average queue length. α and β are coefficient of $t_{average}$ and $l_{average}$, respectively.

The pseudo-code for CMA-ES is given below:

Algorithm CMA-ES.

Step 1: Set $\lambda =$ total number of samples per iteration
and initialize state variables.
Step 2: While

For $k = 1 : \lambda$

$$x_k^{(g+1)} \sim N(x_\mu^g, \sigma^{(g)^2} C^{(g)})$$

Update fitness
Update $ceil(\log(|10 \times \lambda \times N(0, 1)|))$
Sort the individual outputs
Update $x_W^{(g+1)}$, as $x_W^{(g+1)} = \sum_{i=1}^{\mu} w_i x_{i:\lambda}^{(g)}$
Update $p_c^{(g+1)}$, according to Eq. (12.4)
Update $C^{(g+1)}$, according to Eq. (12.5)
Update $p_\sigma^{(g+1)}$, according to Eq. (12.6)
Update $\sigma^{(g+1)}$, according to Eq. (12.7)

12.4 Experiment and Results

In this section, we present our experiment simulation environment. We then describe the details of the parallel reinforcement framework.

12.4.1 Experiment Setup

All the experiments are performed in the Simulation of Urban Mobility (SUMO) tool, which is a well-known open-source traffic simulator and provides various Application Programming Interfaces (APIs). In particular, we utilized SUMO v0.32.0, and the intersection used in this experiment is shown in Fig. 12.1. Approach to the intersection is two lanes and one left turn bay.

The arrival rates of four lanes are random in each episode, while the sum of them will be constant for maintaining a demand baseline. In each episode, random number seeds will also be changed to guarantee the randomness in every simulation.

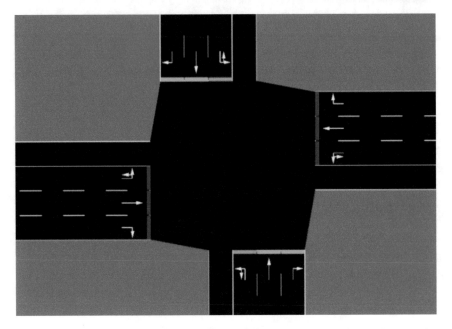

Fig. 12.1 The intersection layout for the traffic simulation

The parallel computing platform we used is HT-condor, which could allocate computing resources based on computing tasks [1]. As the result, we could perform several hundred simulators at the same time.

12.4.2 Optimization of Traffic Signal Parameters

We optimize the parameters mentioned in Table 12.1. using the Covariance Matrix Adaption Evolution Strategy (CMA-ES) algorithm, which is a policy search algorithm, and it has the ability to successively generate and evaluate sets of candidates sampled from a multivariate Gaussian distribution. As shown in Fig. 12.2. once a group of candidates are generated, each one in the group is evaluated with respect to a fitness measure, and after all the candidates are evaluated, the mean of the multivariate Gaussian distribution is recalculated as a weighted average of the candidates with the optimal fitness, and the covariance of which is updated in the direction of previously successful search steps. As CMA-ES belongs to parallel search algorithm, we had the ability to automate and parallelize the learning in high-throughout computer cluster, and this helps us accelerate the optimization which requiring 100000 evaluations in less than ten hours, and this is roughly a 200 times speedup over optimizing without parallel running which would have taken over several months to complete.

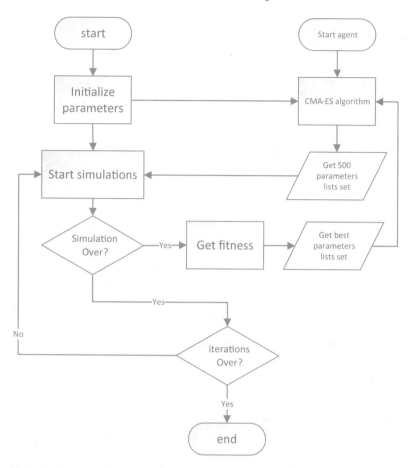

Fig. 12.2 The flow chat of optimization mechanism

12.4.3   Simulation Results

To evaluate the performance of the method we propose in this paper, a baseline traffic controller which using Webster method [16] is compared. We ran the SUMO simulator using the configuration setting explained in Sect. 12.4.1 for the proposed model and the fitness value, average queue length and average delay achieved were compared to the baseline. Figure 12.2 shows the performance of max and average fitness received during the evaluation time for the method proposed, as shown in the diagram, the method proposed performs significantly better than the base line and obtains more reward by iterating, which reflects that the CMA-ES algorithm having the ability to learn an optimal control policy smoothly (Fig. 12.3).

Figure 12.4 illustrates average cumulative delay and average queue length of the intersection using the CMA-ES algorithm and the baseline during the evaluation time,

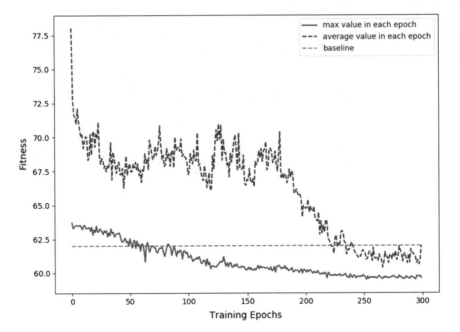

Fig. 12.3 The performance of max and average fitness received during the evaluation time for the method proposed

which are two of the most common performance metrics in the traffic signal control literature. With the iterating of CMA-ES algorithm, the diagram clearly shows that the CMA-ES algorithm is able to find an optimal solution decreasing queue length and total cumulative delay comparing against fixed time method [16]. Figure 12.4 also shows that the average cumulative delay is reduced by 40% and the average queue length of intersection has a 35% discount comparing with fixed time policy.

12.5 Conclusions

In this paper, we applied a parallel reinforcement learning method to traffic signal timing problem in order to find an optimal timing solution, by using traffic simulator. The proposed method can improve the performance by 35–40% compared to fix timing plan.

Future work will keep focus on applying parallel reinforcement learning method to multi-intersection and connected vehicles [17–22], which is much more complex. There are several methods designed to solve it, such as the combination of layered learning and parallel reinforcement learning, deep reinforcement learning and so on. How to apply reinforcement learning method to traffic signal timing of multi-intersection is one of the hottest topics in optimization problems.

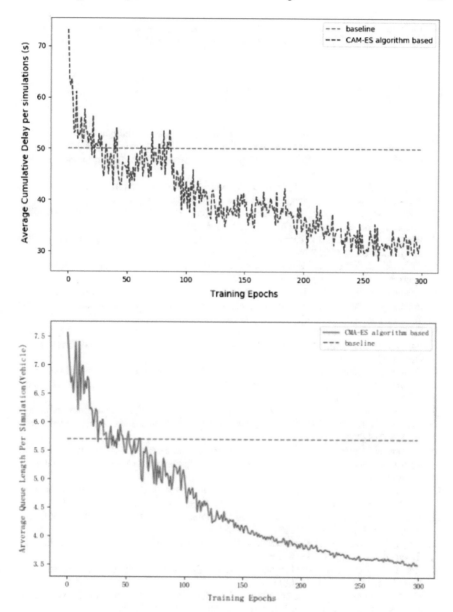

Fig. 12.4 Average cumulative delay and average queue length of the intersection using the method proposed and the baseline during the evaluation time

Acknowledgements This work was supported by the Zhejiang Provincial Natural Science Foundation of China (LQ17E080001), and the China Postdoctoral Science Foundation.

References

1. Mannion, P., Duggan, J., Howley, E.: An experimental review of reinforcement learning algorithms for adaptive traffic signal control. In: McCluskey, L.T., Kotsialos, Kotsialos, A., Muller, P.J., Kluegl, F., Rana, O., Schumann, R. (Eds.), Autonomic Road Transport Support Systems. Springer, Cham, pp. 47–66. (2016)
2. Araghi, S., Khosravi, A., Creighton, D.: A review on computational intelligence methods for controlling traffic signal timing. Expert Syst. Appl. **42**(3), 1538–1550 (2015)
3. Wiering, M.: Multi-agent reinforcement learning for traffic light control. In: 17th International Conference on Machine Learning, pp. 1151–1158 (2000)
4. Wiering, M., Vreeken, J., Veenen, J.V., Koopman, A: Simulation and optimization of traffic in a city. IEEE Intell. Veh. Symp., 453–458 (2004)
5. Abdulhai, B., Kattan, L.: Reinforcement learning: introduction to theory and potential for transport applications. Can. J. Civ. Eng. **30**(6), 981–991 (2003)
6. Aslani, M., Mesgari, M.S., Wiering, M.: Adaptive traffic signal control with actor-critic methods in a real-world traffic network with different traffic disruption events. Transp. Res. Part C Emerg. Technol. **85**(September), 732–752 (2017)
7. El-Tantawy, S., Abdulhai, B., Abdelgawad, H.: Design of reinforcement learning parameters for seamless application of adaptive traffic signal control. J. Intell. Transp. Syst. Technol. Plan. Oper. **18**(3), 227–245 (2014)
8. Darmoul, S., Elkosantini, S., Louati, A., Said, L.B.: Multi-agent immune networks to control interrupted flow at signalized intersections. Transp. Res. Part C: Emerg. Technol. **82** (Suppl. C), 290–313 (2017)
9. Deepeka Garg, G.V., Maria, C.: Deep reinforcement learning for autonomous traffic light control. In: 2018 3rd IEEE International Conference on Intelligent Transportation Engineering, pp. 214–218 (2018)
10. Hansen, N., Ostermeier, A.: Adapting arbitrary normal mutation distributions in evolution strategies: the covariance matrix adaptation. In: IEEE International Conference on Evolutionary Computation, pp. 312–7 (1996)
11. Reddy, S.S.: Energy and spinning reserve scheduling for a wind-thermal power system using CMA-ES with mean learning technique. Electr. Power Energy Syst. 113–122 (2013)
12. Lee, T.Y.: Optimal spinning reserve for a wind-thermal power system using EIPSO. IEEE Trans Power Syst. **22,**1612–1621, (2007)
13. Auger, A., Hansen, N.: A restart CMA evolution strategy with increasing population size. IEEE Congr Evol Comput, 1769–76 (2005)
14. Hansen, N., Ostermeier, A.: Completely de randomized self-adaptation in evolution strategies. Evol. Comput. **9**, 159–195 (2001)
15. Hanse, N.: Reducing the time complexity of the derandomized evolution strategy with covariance matrix adaptation (CMA-ES). Evol. Comput. **11**(1), 1–18 (2003)
16. Webster, F.V.: Traffic Signal Settings. Road Research Technical Paper No.39, Department of Scientific and Industrial Road Research Laboratory. Her Majesty's Stationery Office. London, England (1958)
17. Ma, D., Luo, X., Jin, S., Guo, W., Wang, D.: Estimating maximum queue length for traffic lane groups using travel times from video-imaging data. IEEE Intell. Transp. Syst. Mag. **10**, 123–134 (2018)
18. Li, X., Ghiasi, A., Xu, Z., Qu, X.: A piecewise trajectory optimization model for connected automated vehicles: exact optimization algorithm and queue propagation analysis. Transp. Res. Part B **118**, 429–456 (2018)

19. Ma, D., Luo, X., Wang, D., Jin, S., Wang, F., Guo, W.: Lane-based saturation degree estimation at signalized intersections using travel time data. IEEE Intell. Transp. Syst. Mag. **9**, 136–148 (2017)
20. Li, X., Medal, H., Qu, X.: Connected heterogeneous infrastructure location design under additive service utilities. Transp. Res. Part B **120**, 99–124 (2019)
21. Xu, Z., Wei, T., Easa, S., Zhao, X., Qu, X.: Modeling relationship between truck fuel consumption and driving behavior using data from internet of vehicles. Comput.-Aided Civ. Infrastruct. Eng. **33**, 209–219 (2018)
22. Jin, S., Luo, X., Ma, D.: Determining the breakpoints of fundamental diagrams. IEEE Intell. Transp. Syst. Mag. **11**, 110–120 (2019)
23. Hansen N. The CMA evolution strategy: a comparing review, towards a new evolutionary computation. In: Advances in Estimation of Distribution Algorithms. Springer, Berlin (2006)
24. Raman, R., Livny, M., Solomon, M.: Gang-matching: Advanced resource management through multilateral matchmaking. In Proceedings of the Ninth IEEE International Symposium on High Performance Distributed Computing (2000)

Chapter 13
The Influences of Preceding Vehicle Taillights on Macro Traffic Flow

Jian Zhang, Meng-Tian Li, Tie-Qiao Tang and Yun Zou

Abstract Previous studies have shown that preceding vehicle taillights can have influences on driving behavior. In this paper, we present a macroscopic traffic flow model that considers the front taillights to investigate the impacts of taillights on the evolutions of traffic flow in three typical states (i.e., shock waves, rarefaction waves, and small perturbations). The numerical results indicate that under the influence of preceding vehicle taillights the shock wave is smoother and the rarefaction wave is unchanged. The small perturbation can dissipate without amplification at moderate traffic flow density when the driver's sensitivity to the taillights of the preceding vehicle is sufficiently large. The results indicate that more attention to the preceding vehicle taillights can enhance traffic safety and stability.

13.1 Introduction

Traffic flow modeling and simulation is a convenient, inexpensive and effective method to study traffic problems (e.g., jams, accidents, pollution, etc.). Various traffic models are proposed by researchers to investigate the formation mechanisms of traffic problems and to help design control strategies. These models can be simply categorized into macro and micro models.

J. Zhang · T.-Q. Tang (✉)
School of Transportation Science and Engineering, Beijing Key Laboratory for Cooperative Vehicle Infrastructure Systems and Safety Control, Beihang University, Beijing 100191, China
e-mail: tieqiaotang@buaa.edu.cn

M.-T. Li
Research Center of Geotechnical and Structural Engineering, Shandong University, Jinan 250061, China

J. Zhang · M.-T. Li
Department of Architecture and Civil Engineering, Chalmers University of Technology, SE412 96 Gothenburg, Sweden

Y. Zou
School of Civil and Environmental Engineering, University of Technology Sydney, Sydney, NSW 2007, Australia

© Springer Nature Singapore Pte Ltd. 2019
X. Qu et al. (eds.), *Smart Transportation Systems 2019*, Smart Innovation,
Systems and Technologies 149, https://doi.org/10.1007/978-981-13-8683-1_13

The macroscopic traffic flow models are mainly used to study the distribution, evolution, and propagation of traffic flow by means of some macro variables, such as flow, density, and velocity. Among these macro models, the first simple but effective model was independently developed by Lighthill and Whitham [1] and Richards [2] known as LWR model. The LWR model can reproduce the formation and evolution of shock wave well, but it cannot be used to study the traffic flow in nonequilibrium state because the speed cannot deviate from the equilibrium speed. To overcome this drawback, the equilibrium speed was substituted by acceleration and many high-order models [3–5] are proposed by researchers. The high-order models can be divided into the density-gradient (DG) models and speed-gradient (SG) models. The Payen model [3] is a representative DG model but have the problem that there exists a characteristic speed that is greater than the macroscopic flow velocity. In order to eliminate this problem, Jiang et al. [4] developed a SG model by introducing full velocity difference model [6] and applying the connection method of macro–micro variables. A few other extended macro traffic flow models were proposed hereafter to investigate the influences of some certain factors on macro traffic flow. The microscopic traffic models, also known as the car-following models, are used to explore driving behavior and corresponding complex traffic phenomena through some microvariables (e.g., vehicle acceleration, speed, and headway) [6–8].

Models [1–8] can depict many traffic phenomena and explain their formation mechanisms, but these models cannot be used to directly study the effects of preceding vehicle taillights on traffic flow since they did not explicitly consider this factor. In fact, the taillights of preceding car have a remarkable influence on traffic flow. Nagatani [9] investigated the chain-reaction crashes in traffic flow affected by taillights. Zhang et al. [10] studied how the preceding car's taillight influences the driving behavior, and proposed a car-following model to investigate the impacts of preceding vehicle taillights on each car's driving behavior. Numerical results show that preceding car's taillights can improve the stability and safety of traffic. However, Zhang et al. [10] did not explore the impacts from the macroscopic perspective. In this paper, we convert the micro variables [10] into macro variables and develop a macro traffic flow model to explore the effects of preceding vehicle taillights on traffic flow. This paper is organized as follows: the traffic flow model with the preceding vehicle taillights is developed in Sect. 13.2, numerical experiments are carried out to explore the influences of preceding vehicle taillights on traffic flow in Sects. 13.3, and 13.4 summarizes the conclusions.

13.2 Macro Traffic Model

The proposed macro model is based on SG model, the simplest SG model can be formulated as follows [4]:

$$\begin{cases} \rho_t + (\rho v_e(\rho))_x = 0 \\ v_t + v v_x = \frac{v_e(\rho) - v}{T} + c_0 v_x \end{cases}, \tag{13.1}$$

where ρ is the traffic density; $v_e(\rho)$ is the equilibrium speed at density ρ; c_0 is the propagation speed of small perturbation; T is the relaxation time; x and t represent space and time, respectively.

The model does not consider preceding vehicle taillights so it cannot be used to study the effects of front taillights on macro traffic flow. In order to study the impacts of preceding vehicle taillights on driving behavior, Zhang et al. [10] developed a car-following model that takes into account the taillights of preceding vehicle, i.e.,

$$\frac{dv_n(t)}{dt} = f(v_n(t), \Delta x_n(t), \Delta v_n(t), \varepsilon_n(t)) + \theta_n(t)\xi_n(t), \tag{13.2}$$

where $v_n(t)$, $\Delta x_n(t)$, $\Delta v_n(t)$ are respectively the nth car's speed, spatial headway and relative speed with the preceding car; $\varepsilon_n(t)$ is the collection of other factors that influence the nth car's driving behavior; $\theta_n(t)$ is a binary variable that reflects whether the $(n-1)$th car's taillights are red, i.e.,

$$\theta_n(t) = \begin{cases} 0, & \text{if } a_{n-1}(t) \geq 0 \\ 1, & \text{otherwise} \end{cases}. \tag{13.3}$$

$\xi_n(t)$ is the influenced term produced by the $(n-1)$th car's taillights, which can be formulated as follows:

$$\xi_n(t) = \begin{cases} \xi_0 \tanh\left(1 - \frac{\Delta x_n(t)}{\Delta x_0}\right) \times H(-\Delta v_n(t)) \times \Delta v_n(t), & \text{if } \Delta x_n(t) \leq \Delta x_0 \\ 0, & \text{otherwise} \end{cases}, \tag{13.4}$$

where ξ_0 is a sensitivity parameter; H is the Heaviside function, which is defined as follows:

$$H(x) = \begin{cases} 1, & \text{if } x \geq 0 \\ 0, & \text{otherwise} \end{cases}. \tag{13.5}$$

Applying the same method [4], we can propose a macroscopic traffic flow model that considers preceding vehicle taillights, where the control equation can be formulated as follows:

$$\begin{cases} \rho_t + (\rho v_e(\rho))_x = 0 \\ v_t + vv_x = \frac{v_e(\rho) - v}{T} + c_0 v_x + \theta\xi \end{cases}, \tag{13.6}$$

where θ is the indicator parameter of preceding vehicle taillight which is the same as Eq. (13.3), ξ is the macro influence term of front taillights, similar to Eq. (13.4), it can be formulated as follows:

$$\xi = \begin{cases} \xi_0 \tanh\left(1 - \frac{\rho^*}{\rho}\right) \times H(-\Delta v) \times \Delta v, & \text{if } \rho \geq \rho^* \\ 0, & \text{otherwise} \end{cases}, \qquad (13.7)$$

where ρ^* is a specific density, when the density is greater than ρ^*, the taillights of the preceding vehicle will affect the subsequent traffic flow.

By introducing the car-following model considering the front taillights and applying usual micro–macro variables transformation method, the macro traffic flow model with consideration of preceding vehicle taillights is developed. Compared with other traffic models, our model can be directly used to study the impact of taillights on macro traffic flow. In addition, by changing the value of ξ_0 in Eq. (13.7), the influences of sensitivity to taillight on traffic flow can also be investigated from the macroscopic perspective.

13.3 Numerical Experiments

In this section, we conduct some numerical experiments to investigate the phenomena that the proposed model can reproduce, and to analyze the influences of preceding vehicle taillights on macro traffic flow under shock wave, rarefaction wave, and small disturbance. Since it is difficult to obtain the analytical solution of Eq. (13.6), we utilize the one-order upwind difference to discretize Eq. (13.6), i.e.,

$$\rho_i^{j+1} = \rho_i^j + \frac{\Delta t}{\Delta x} \rho_i^j \left(v_i^j - v_{i+1}^j\right) + \frac{\Delta t}{\Delta x} v_i^j \left(\rho_{i-1}^j - \rho_i^j\right), \qquad (13.8)$$

if $v_i^j \leq c_0$,

$$v_i^{j+1} = v_i^j + \frac{\Delta t}{\Delta x}\left(c_0 - v_i^j\right)\left(v_{i+1}^j - v_i^j\right) + \Delta t \theta_i^j \xi_i^j + \frac{\Delta t}{T}\left(v_e\left(\rho_i^j\right) - v_i^j\right), \qquad (13.9)$$

and otherwise,

$$v_i^{j+1} = v_i^j + \frac{\Delta t}{\Delta x}\left(c_0 - v_i^j\right)\left(v_i^j - v_{i-1}^j\right) + \Delta t \theta_i^j \xi_i^j + \frac{\Delta t}{T}\left(v_e\left(\rho_i^j\right) - v_i^j\right), \qquad (13.10)$$

where i, j are spatial and temporal indexes, respectively; Δx is the space step; Δt is the time step.

First, we study the effects of preceding vehicle taillights on the evolutions of shock wave and rarefaction wave under the free boundary condition. The initial conditions are as follows:

$$\begin{cases} \rho_{up}^a = 0.04 \text{ veh/m}, \ \rho_{down}^a = 0.18 \text{ veh/m} \\ \rho_{up}^b = 0.18 \text{ veh/m}, \ \rho_{down}^b = 0.04 \text{ veh/m} \end{cases}, \qquad (13.11)$$

where ρ_{up}^i, ρ_{down}^i ($i = a, b$) are the initial density at the upstream and downstream, respectively. The equilibrium speed–density relationship proposed by Del Castillo and Benitez is applied [4]:

$$v_e(\rho) = v_f\left[1 - \exp\left(1 - \exp\left(\frac{c_m}{v_f}\left(\frac{\rho_j}{\rho} - 1\right)\right)\right)\right],\qquad(13.12)$$

where v_f is the free-flow speed; ρ_j is the jamming density; c_m is the kinematic wave speed at the jamming density. Here, the road length is 20 km, the discontinuous point of the density is the midpoint of the road, the discrete space step is 200 m and the time step is 1 s. The initial speed conditions are as follows:

$$v_{up}^{a,b} = v_e\left(\rho_{up}^{a,b}\right),\quad v_{down}^{a,b} = v_e\left(\rho_{down}^{a,b}\right).\qquad(13.13)$$

Other related parameters are as follows:

$$T = 10\ \text{s},\ v_f = 30\ \text{m/s},\ c_0 = c_m = 11\ \text{m/s},\ \rho_j = 0.2\ \text{veh/m},\ \rho^* = 0.01\ \text{veh/m}.\qquad(13.14)$$

And the sensitivity parameter ξ_0 is set to 0, 0.1, 0.2, and 0.5 to study the influences of sensitivity to front taillights on the evolution of traffic flow.

Figure 13.1 displays the density evolutions of shock wave under different ξ_0. The results show that preceding vehicle taillights can smooth the shock wave and the shock wave is smoother with the increase of ξ_0. Figure 13.2 shows the density evolution of rarefaction wave. Regardless of the value of ξ_0, the evolution patterns are the same, which indicates that the taillights of the preceding vehicle have no impact on the rarefaction waves. A shock wave situation corresponds to a situation where nearly free-flow traffic encounters jamming traffic and the free-flow traffic must decelerate to avoid crashes. A rarefaction wave situation corresponds to a situation where jam traffic accelerates to a nearly free-flow speed. It is obvious that the taillights could only affect the traffic due to the deceleration process, so the shock waves are changed with the influence of taillight whereas the rarefaction waves are not changed. The smoother changes of shock waves suggest that the situations of sharp deceleration are reduced influenced by the taillights which improves the traffic safety and comfort. And if we pay more attention to the taillight of preceding traffic, the improvement will be increased.

Next, we study the evolution of a small disturbance under the proposed model. A localized perturbation in a homogeneous traffic flow is set as the initial condition which can be formulated as follows [4]:

$$\rho(x, 0) = \rho_0 + \Delta\rho\left(\cosh^{-2}\left(\frac{160}{L}\left(x - \frac{5L}{16}\right)\right) - \frac{1}{4}\cosh^{-2}\left(\frac{40}{L}\left(x - \frac{11L}{32}\right)\right)\right),\qquad(13.15)$$

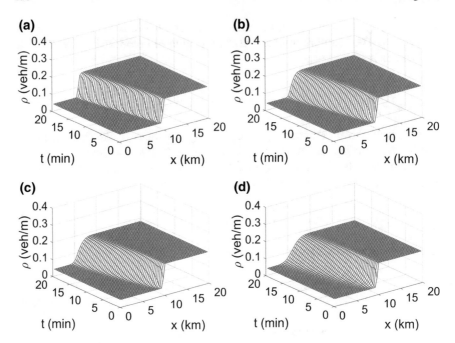

Fig. 13.1 The density evolutions of shock wave, where ξ_0 in (**a**)–(**d**) are, respectively, 0, 0.1, 0.2, and 0.5

Fig. 13.2 The density evolution of rarefaction wave

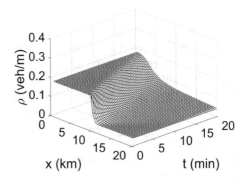

where L = 32.2 km is the road length, the discrete space step is 100 m and time step is 1 s, periodic boundary conditions are adopted.

We apply the equilibrium speed–density relationship proposed by Kerner and Konhäuser in the following simulations, which is the same as in Ref. [4]:

$$v_e(\rho) = v_f\left(1/\left(1 + \exp\left(\frac{\rho/\rho_j - 0.25}{0.06}\right)\right) - 3.72 \times 10^{-6}\right). \tag{13.16}$$

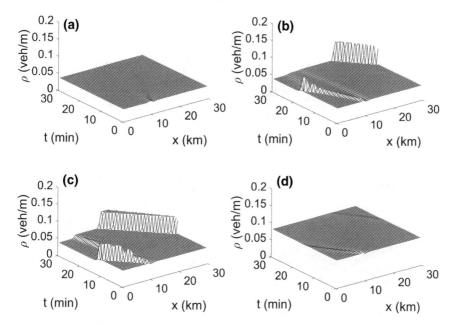

Fig. 13.3 The density evolution of the small perturbation when $\xi_0 = 0.1$, where ρ_0 in (**a**)–(**d**) are, respectively, 0.035, 0.042, 0.046, and 0.08

The initial speed is equal to the equilibrium speed everywhere, i.e., $v(x, 0) = v_e(\rho(x, 0))$, and $\Delta\rho_0 = 0.01$ veh/m. The sensitivity parameter ξ_0 is set to 0.1 and 0.5, and the other parameters are the same as those in the above simulations of shock and rarefaction wave.

Figures 13.3 and 13.4 display the traffic density evolutions of the small perturbation under different ρ_0 and ξ_0. From these figures, we can find the following results: When ρ_0 is very low or high, the preceding vehicle taillights have no prominent effects on the evolution of small perturbation regardless of the value of the sensitivity parameter since the traffic flow is very stable. At this time, the small perturbation can quickly evolve into the uniform flow.

When ρ_0 is moderate, the traffic flow is unstable. If we do not consider the impact of preceding vehicle taillights, the small disturbance will evolve into stop-and-go traffic. The taillights have prominent effects on the evolution of small perturbation and the degree of influence is related to the sensitivity coefficient. When ξ_0 is relatively low, the stop-and-go traffic appears but the amplitude is smaller than the situation in Ref. [4]. The amplitude of the stop-and-go is decreased with the increase of ξ_0. When the value of ξ_0 is relatively high, i.e. 0.5, the stop-and-go is disappeared and the small perturbation can be dissipated without any amplification. The numerical results show that the taillight of preceding vehicle can improve the stability of traffic flow.

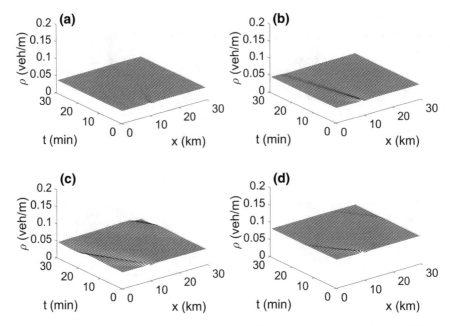

Fig. 13.4 The density evolution of the small perturbation when $\xi_0 = 0.5$, where ρ_0 in (**a**)–(**d**) are, respectively, 0.035, 0.042, 0.046, and 0.08

13.4 Conclusion

Although many macro traffic flow models have been developed to explore various complex traffic phenomena, they cannot be used to study the impacts of preceding vehicle taillights on traffic flow, as this factor is not directly considered. In this paper, we transform the micro variables in Ref. [10] into the macro ones, and develop a macro traffic flow model to explore the effects of preceding vehicle taillights on traffic flow in some typical situations. The numerical results show that the proposed model can perfectly describe the complex traffic phenomena influenced by front taillights. The preceding vehicle taillights can smooth the shock wave and have no effect on the rarefaction wave. The stop-and-go phenomenon caused by a small disturbance can be reduced or eliminated. These findings suggest that drivers should pay more attention to the taillights, and some intelligent driving assistance devices should be developed to improve traffic safety and stability based on the influences of preceding vehicle taillights on traffic flow.

However, the relevant parameters and numerical results are not testified by empirical or experimental data. In view of the limitations of this paper, we should apply some empirical or experimental data to propose a more realistic model [11–13] in the future and further study the effects of preceding vehicle taillights on the evolution of traffic flow.

Acknowledgements This work was supported by the National Natural Science Foundation of China (71422001, 71271016).

References

1. Lighthill, M.J., Whitham, G.B.: On kinematic waves II. A theory of traffic flow on long crowded roads. Proc. R. Soc. Lond. Ser. A **229**, 317–345 (1955)
2. Richards, P.I.: Shock waves on the highway. Oper. Res. **4**, 42–51 (1956)
3. Payne, H.J.: Models of Freeway Traffic and Control. Mathematical Models of Public System, pp. 51–61 (1971)
4. Jiang, R., Wu, Q.S., Zhu, Z.J.: A new continuum model for traffic flow and numerical tests. Transp. Res. Part B **36**, 405–419 (2002)
5. Tang, T.Q., Caccetta, L., Wu, Y.H., Huang, H.J., Yang, X.B.: A macro model for traffic flow on road networks with varying road conditions. J. Adv. Transp. **48**, 304–317 (2014)
6. Jiang, R., Wu, Q.S., Zhu, Z.J.: Full velocity difference model for a car-following theory. Phys. Rev. E **64**, 017101 (2001)
7. Zhou, M., Qu, X., Li, X.: A recurrent neural network based microscopic car following model to predict traffic oscillation. Transp. Res. Part C **84**, 245–264 (2017)
8. Zhou, M., Qu, X., Jin, S.: On the impact of cooperative autonomous vehicles in improving freeway merging: a modified intelligent driver model based approach. IEEE Trans. Intell. Transp. Syst. **18**, 1422–1428 (2017)
9. Nagatani, T.: Chain-reaction crash in traffic flow controlled by taillights. Phys. A **419**, 1–6 (2015)
10. Zhang, J., Tang, T.Q., Yu, S.W.: An improved car-following model accounting for the preceding car's taillight. Phys. A **492**, 1831–1837 (2018)
11. Qu, X., Wang, S., Zhang, J.: On the fundamental diagram for freeway traffic: A novel calibration approach for single-regime models. Transp. Res. Part B **73**, 91–102 (2015)
12. Qu, X., Zhang, J., Wang, S.: On the stochastic fundamental diagram for freeway traffic: Model development, analytical properties, validation, and extensive applications. Transp. Res. Part B **104**, 256–271 (2017)
13. Zhang, J., Qu, X., Wang, S.: Reproducible generation of experimental data sample for calibrating traffic flow fundamental diagram. Transp. Res. Part A **111**, 41–52 (2018)

Chapter 14
Towards Eliminating Overreacted Vehicular Maneuvers: Part I Model Development and Calibration

Yang Yu, Yun Zou and Xiaobo Qu

Abstract Microscopic car following models are of great importance to traffic flow studies and vehicular dynamics reproducing. The Full Velocity Difference (FVD) model is a well-known example with satisfactory simulation performances in most times. However, by analyzing the structure of the model formulas, we find that it can sometimes generate overreacted vehicular maneuvers such as unrealistically strong (overshooting for short) accelerations or decelerations that conflict with normal driver habits or even beyond the actual vehicular acceleration/deceleration performance, especially when the target vehicle encounter a leader cut-in or move out (leader lane change for short). As Part I of the entire research, this paper corrects the above deficiency of the FVD model by proposing a capped-Full Velocity Difference (capped-FVD) model in which we limit any potential overshooting accelerations or decelerations generated to a reasonable range. Then, all model parameters are also calibrated using field data. Performance comparative analyses to validate the performance improvement of the capped-FVD model are included in the other companion paper serving as Part II of this research.

14.1 Introduction

Microscopic car following models are used to reproduce how vehicles are following one another on real roadways. They are the foundations of microscopic traffic flow theories and is the most important tool in traffic modelling. While some of the latest

Y. Yu (✉) · Y. Zou
School of Civil and Environmental Engineering, University of Technology Sydney, Ultimo, NSW 2007, Australia
e-mail: Yang.Yu-4@student.uts.edu.au

Y. Zou
e-mail: Yun.Zou@student.uts.edu.au

X. Qu
Department of Architecture and Civil Engineering, Chalmers University of Technology, 412 96 Gothenburg, Sweden
e-mail: xiaobo@chalmers.se

© Springer Nature Singapore Pte Ltd. 2019
X. Qu et al. (eds.), *Smart Transportation Systems 2019*, Smart Innovation, Systems and Technologies 149, https://doi.org/10.1007/978-981-13-8683-1_14

car following models and algorithms [1, 2] dedicatedly focus on simulating connected automated vehicles, traditional car following models mainly aims at reproducing behaviours of manual vehicles and are normally formula-based that take a vehicle's current status and surrounding traffic conditions as inputs and output the movement of the vehicle at next time step mainly in the form of acceleration rate since acceleration is directly linked to control of engine power [3]. Two early examples include the Gazis–Herman–Rothery (GHR) model [4, 5] and the Linear (Helly) model [6], which were followed by some famous descendants such as the Gipps model [7], the Optimal Velocity (OV) model [8], and the Intelligent Driver Model (IDM) family [9–11]. Among these well-known traditional car following models, the OV model has the simplest model structure yet can still reproduce vehicular dynamics of normal drivers. The OV model simply considers the impact of the immediate leader and assume that a driver will always identify an optimal velocity based on the space headway (gap) to the immediate leader and attempt to apply different degrees of accelerations according to the velocity difference between the gap-dependent optimal velocity and current actual velocity. The OV model was later improved to a Generalized Force (GF) model by Helbing and Tilch [12] through introducing another term into the OV function which represents the impact of negative velocity difference between the leader and the follower. Traffic simulation result shows that the GF model can less frequently produce too strong accelerations or decelerations but better avoid rear-end collisions than the OV model. Then, the GF model was further upgraded to a Full Velocity Difference (FVD) model by Jiang et al. [13] to address the unsatisfactory performance of the existing model in reproducing both the delay time of car motion and kinematic wave speed at jam density. Compared with the GF model, the FVD model takes not only the negative leader–follower velocity difference, but also the positive velocity difference into consideration.

Although the FVD model is already a highly improved one in the OV model family, we find that it may produce overshooting accelerations or decelerations that conflict with normal driver habits (Normal drivers will hardly apply accelerations stronger than 3.41 m/s^2 or decelerations stronger than -3.41 m/s^2 [14, 15]) or even beyond the actual vehicular acceleration/deceleration performance (4.50 m/s^2 is the typical maximum acceleration a modern vehicle can reach [16]; -8.41 m/s^2 is the typical maximum deceleration a modern vehicle with anti-lock braking system can reach upon the speed of 60 km/h [17]) in certain traffic scenarios, especially when the target follower encounters leader lane change. In other words, the FVD model may overreact in such traffic scenarios. Thus, the FVD model cannot be applied in multi-lane scenarios where lane changes occur frequently. To address this, we propose a capped-FVD model based on the existing FVD model by introducing an acceleration cap and a deceleration cap, respectively, and limit any overshooting accelerations/decelerations outside the acceleration/deceleration cap to smaller but more reasonable values. By doing so, the capped-FVD model is expected to be able to eliminate overacted vehicular maneuvers and better describe vehicular behaviours in aforementioned traffic scenarios where overshooting accelerations or decelerations may be yielded by the FVD model, and still has the same good performance as the existing FVD model in other normal traffic scenarios.

The rest of the paper (Part I of the research) is organized as follows: Sect. 14.2 first introduces the structure of the FVD model and point out its potential deficiency. Then, a capped-FVD model is proposed to address the deficiency. Section 14.3 first describes the traffic data adopted for model calibration. Then, all the model parameters are calibrated based on these data. Section 14.4 shows a conclusion of this paper along with a brief introduction about Part II of this research.

14.2   Model Development

14.2.1   The Full Velocity Difference (FVD) Model and Its Deficiency

As introduced above, the FVD model takes the actual velocity of the follower, the gap-dependent optimal velocity of the follower, and the velocity difference between the leader and the follower (approaching rate) as input and yield the acceleration rate of the follower as output. The corresponding function is listed as follows:

$$\ddot{x}_n(t) = \frac{1}{\tau}[V(\Delta x) - \dot{x}_n(t)] + \lambda \Delta v \tag{14.1}$$

where n refers to the follower and x_n is the its position at time t. The dot $\cdot$ describes the differentiation with respect to time. $V(\Delta x)$ is the optimal velocity of the follower and it is decided by the space headway Δx $(x_{n-1}(t) - x_n(t))$. Parameter τ is the relaxation time which describes the adaptation of the follower to a new optimal velocity as a result of changes in space headway Δx and velocity $\dot{x}_n(t)$ [18]. Besides, approaching rate Δv is calculated by $\dot{x}_{n-1}(t) - \dot{x}_n(t)$ and parameter λ indicates the follower's response sensitivity to Δv.

At last, a general form of the optimal velocity function $V(\Delta x)$ is determined as follows based on the works of Helbing and Tilch [12] and Kesting and Treiber [18]:

$$V(\Delta x) = V_1 + V_2 \tanh\left[\frac{(\Delta x - l)}{l_{int}} - \beta\right] \tag{14.2}$$

where V_1 and V_2 are both velocities and $V_1 + V_2$ depicts the follower's desired velocity in free-flow traffic. Besides, the interaction length l_{int} and the unitless parameter β together define the shape of the equilibrium fundamental diagram [18].

As can be observed from Eqs. (14.1) and (14.2), the FVD model takes more inputs (Δx, $\dot{x}_n(t)$, and Δv) to make predictions compared with both OV model (only Δx and $\dot{x}_n(t)$) and GF model (only Δx, $\dot{x}_n(t)$, and negative Δv), which is also the main reason why the FVD model can outperform the other two in reproducing car following movements. However, it is easy to tell from Eq. (14.2) that the follower optimal velocity $V(\Delta x)$ is monotonically increasing with the increase of Δx. Thus, if both the space headway Δx and approaching rate Δv mutate simultaneously with the same tendency (suddenly increase or decrease by a large margin), both terms in the right-hand side of Eq. (14.1) will also mutate simultaneously with the same tendency, which is very likely to result in the generation of overshooting accelerations or decelerations $\ddot{x}_n(t)$. In other words, the FVD model may overreact in traffic conditions where both space headway Δx and approaching rate Δv simultaneously mutate with the same tendency, which includes but not limited to the following:

1. The current leader moves out to adjacent lane (Δx mutates by suddenly increasing) and the new leader ahead has higher velocity than the current leader (Δv mutates by suddenly increasing);
2. A new leader cuts in from adjacent lane (Δx mutates by suddenly decreasing) with a smaller velocity than the current leader (Δv mutates by suddenly decreasing).

The above two scenarios are not rare in multi-lane roadways and it is clear that the existing FVD model cannot nicely reproduce vehicular dynamics of following vehicles in such traffic conditions due to possible overreacted vehicular maneuvers (overshooting accelerations or decelerations) mentioned above.

14.2.2 The Capped-Full Velocity Difference (Capped-FVD) Model

In order to eliminate the deficiency of the existing FVD model in producing over-reacted vehicular maneuvers (overshooting accelerations or decelerations) in multi-lane traffic scenarios, we propose a capped-Full Velocity Difference (capped-FVD) model and its primary modification made compared to the existing model is the intro-duction of another function to restrict the overshooting accelerations or decelerations generated by Eq. (14.1) to a reasonable range, which is displayed as follows:

$$
a_n(t) = \begin{cases}
a_{cap} : \ddot{x}_n(t) > a_{cap} \cap t = 1 \\
a_{cap} : \ddot{x}_n(t) > a_{cap} \cap \ddot{x}_n(t-1) \leq a_{cap} \\
a_{cap} : \ddot{x}_n(t) > a_{cap} \cap \ddot{x}_n(t-1) > a_{cap} \cap \frac{\ddot{x}_n(t)}{\ddot{x}_n^+} > 1 \\
a_{cap}\sqrt[r]{\frac{\ddot{x}(t)}{\ddot{x}_n^+}} : \ddot{x}_n(t) > a_{cap} \cap \ddot{x}_n(t-1) > a_{cap} \cap 0 < \frac{\ddot{x}_n(t)}{\ddot{x}_n^+} \leq 1 \\
\ddot{x}_n(t) : \ddot{x}_n(t) \subset [d_{cap}, a_{cap}] \\
d_{cap} : \ddot{x}_n(t) < d_{cap} \cap t = 1 \\
d_{cap} : \ddot{x}_n(t) < d_{cap} \cap \ddot{x}_n(t-1) \geq d_{cap} \\
d_{cap} : \ddot{x}_n < d_{cap} \cap \ddot{x}_n(t-1) < d_{cap} \cap \frac{\ddot{x}_n(t)}{\ddot{x}_n^-} > 1 \\
d_{cap}\sqrt[r]{\frac{\ddot{x}_n(t)}{\ddot{x}_n^-}} : \ddot{x}_n(t) < d_{cap} \cap \ddot{x}_n(t-1) < d_{cap} \cap 0 < \frac{\ddot{x}_n(t)}{\ddot{x}_n^-} \leq 1
\end{cases} \tag{14.3}
$$

where parameter a_{cap} is a positive value standing for a considerably high accelera-tion rate that both main stream vehicles can reach and most drivers would be able to maintain while still safely controlling the vehicle. Similarly, parameter d_{cap} is a negative value standing for a considerably high deceleration rate based on the same criteria. The introduction of both caps not only well matches the actual acceleration and deceleration ability of most vehicles, but also addresses the fact that most drivers would hardly use a too large acceleration or deceleration on roadways for both safety and comfortable driving reasons. $\ddot{x}_n(t)$ still refers to the acceleration or deceleration generated by the existing FVD model (Eq. (14.1)) and if $\ddot{x}_n(t)$ lies outside the range of $[d_{cap}, a_{cap}]$, it is regarded as overshooting. $a_n(t)$ is the final acceleration or deceleration generated by the modified model. r is the degree of radical. Besides, $\ddot{x}_n^+$ (>0) and $\ddot{x}_n^-$ (<0) represent the maximum overshooting acceleration and deceleration generated by Eq. (14.1), respectively. Their values should be updated according to the following rules:

1. Update the value of $\ddot{x}_n^+$ as $\ddot{x}_n(t)$ every time when Eq. (14.1) generates an overshooting acceleration at time t ($\ddot{x}_n(t) > a_{cap}$) but $t = 1$ or it does not generate overshooting acceleration at time $t - 1$ ($\ddot{x}_n(t - 1) \leq a_{cap}$). This usually corresponds to the moment when the immediate leader just moves out;
2. Update the value of $\ddot{x}_n^-$ as $\ddot{x}_n(t)$ every time when Eq. (14.1) generates an overshooting deceleration at time t ($\ddot{x}_n(t) < d_{cap}$) but $t = 1$ or it does not generate overshooting deceleration at time ($t - 1$ ($\ddot{x}_n(t - 1) \geq d_{cap}$). This usually corresponds to the moment when a new leader just cuts in;
3. Update the value of $\ddot{x}_n^+$ as $\ddot{x}_n(t)$ every time when an overshooting $\ddot{x}_n(t)$ larger than the current $\ddot{x}_n^+$ is generated;
4. Update the value of $\ddot{x}_n^-$ as $\ddot{x}_n(t)$ every time when an overshooting $\ddot{x}_n(t)$ smaller than the current $\ddot{x}_n^-$ is generated.

According to Eq. (14.3), whenever an overshooting acceleration $\ddot{x}_n(t)$ that is larger than a_{cap} is generated by Eq. (14.1), it will be reduced down to:

1. a_{cap}, if $\ddot{x}_n(t - 1)$ is smaller than or equal to a_{cap}. This means that whenever the simulated vehicle encounters a new special traffic condition such as a leader move out and generates an overshooting acceleration based on the existing FVD model, the capped-FVD model will reduce it down to a_{cap};
2. a_{cap}, if $\ddot{x}_n(t - 1)$ is larger than a_{cap} and $\ddot{x}_n(t)$ is even larger than the current $\ddot{x}_n^+$. This means that for any subsequent moments when an even larger overshooting acceleration is generated, the capped-FVD model will still reduce it down to the already-high a_{cap};
3. $a_{cap}\sqrt[r]{\dfrac{\ddot{x}_n(t)}{\ddot{x}_n^+}}$, if $\ddot{x}_n(t - 1)$ is larger than a_{cap} and $\ddot{x}_n(t)$ is smaller than the current $\ddot{x}_n^+$. This means that the capped-FVD model will reduce down any subsequent overshooting acceleration $\ddot{x}_n(t)$ that is small than or equal to the current maximum overshooting acceleration $\ddot{x}_n^+$ to a reasonable value below a_{cap} according to the ratio of $\ddot{x}_n(t)$ to $\ddot{x}_n^+$. And the ratio is slightly amplified by a radical to compensate for reducing down the original acceleration value generated by Eq. (14.1).

Any overshooting decelerations yielded by Eq. (14.1) will also be processed in the same manner according to Eq. (14.3).

If no overshooting accelerations/decelerations are yielded by the existing FVD model (Eq. (14.1)), the capped-FVD model will reduce to the FVD model and yield the same output.

Equations (14.1)–(14.3) compose the capped-FVD model and it is easy to conclude that the modified model still adopts the same logic of the existing FVD model to generate accelerations or decelerations but would further confine any overshooting accelerations or decelerations to a reasonable range $-\left[d_{cap}, a_{cap}\right]$ m/s^2. The initial value of acceleration or deceleration generated by Eq. (14.1) stands for the initial willingness of the driver to accelerate or decelerate according to the existing FVD model but they are limited by both the acceleration/deceleration performance of real vehicles and safety/comfortable driving requirements of normal drivers, which is represented by Eq. (14.3). In doing so, the capped-FVD model will no longer yield overacted vehicular maneuvers such as overshooting accelerations or decelerations as was observed in the existing KNN model and the capped-FVD model can better describe vehicular dynamics in traffic scenarios such as leader lane changes.

14.3 Model Calibration

14.3.1 Data Description and Pre-processing

To be consistent with similar studies in the past, we adopt traffic data from NGSIM program [19] to do model calibration and testing in this research. To calibrate all model parameters, we randomly adopt 5 data pairs depicting the movements of 5 different following vehicles and their leaders from either the US Highway 101 (US 101) dataset or the Interstate 80 (I-80) dataset, both of which contain vehicular data collected from multi-lane straight road sections during peak hours (large traffic flows). As such, these leader–follower data pairs are with high quality and can well represent vehicular dynamics on most busy roadways. The data collecting frequency is 0.1 s and each data pair contains the trajectories, velocities and accelerations of both the leader and the follower.

In order to eliminate mutated and false position values (reduce the impact of measurement errors) and smooth the trajectories, we also apply a symmetric exponential moving average filter [20] with a smoothing width of 0.5 s to all the raw trajectory data. Then, the smoothed velocity/acceleration at each time step are calculated from the first/second order differentiations of the corresponding position values. Besides, all the trajectory data chosen start from 2 s before the lane change moment for the convenience of comparison. Finally, we further limit the smoothed accelerations/decelerations to a range of $[-3.41, 3.41]$ m/s^2, which is consistent with both the acceleration/deceleration range in original NGSIM datasets and the deceleration rate used in determining the stopping sight distance [14, 15].

Table 14.1 Calibration results for the FVD model and the capped-FVD model

	τ	λ	V_1	V_2	l_{int}	β	a_{cap}	d_{cap}	r
FVD model	2.94	0.89	14.41	15.89	15.22	1.51			
Capped-FVD model	2.94	0.89	14.41	15.89	15.22	1.51	3.41	−3.41	3

14.3.2   Model Parameter Calibration

Before we can compare the performance of the proposed capped-FVD model and the existing FVD model, we need to calibrate the values of all the model parameters. Given that the capped-FVD model adopts the same formulas as the existing FVD model to generate accelerations/decelerations and the values of its three extra parameters (a_{cap}, d_{cap}, r) are determined based on experiences rather than calibration, we only needs to calibrate the parameters in the existing FVD model. A genetic algorithm proposed by Kesting and Treiber [18] is used to calibrate the model against the calibration dataset. The calibration process aims at minimizing the errors in the space headway Δx, and a mixed error measure objective function [18] is used to measure the above errors quantitatively. To guarantee the rationality of all parameter values, the following parameter restrictions are pre-applied:

1. The relaxation time τ lies in the scope of $[0.1, 4]$ s;
2. The sensitivity parameter λ lies in the scope of $[0.1, 3]$ s^{-1};
3. $V_1 > 0$ m/s, $V_2 > 0$ m/s;
4. The maximum free-flow velocity $V_1 + V_2$ lies in the scope of $[5, 40]$ m/s;
5. The interaction length l_{int} lies in the scope of $[0.1, 80]$ m;
6. The unitless parameter β lies in the scope of $[0.1, 6]$;
7. The average length of vehicles l is 5 m;
8. $V_1 + V_2 \tanh(-\beta) \leq 0$ because every vehicle should at least stop when its net space headway $(\Delta x - l)$ becomes 0 m;
9. The value of a_{cap} is set as 3.41 m/s^2 and the value of d_{cap} is set as −3.41 m/s^2 in order to be consistent with both the acceleration/deceleration range in original NGSIM datasets and the deceleration rate used in determining the stopping sight distance;
10. The value of radical r is set as 3.

The calibration results are displayed in Table 14.1.

14.4   Conclusion

As Part I of this research, a capped-Full Velocity Difference (capped-FVD) car following model is proposed in this paper as a simple variant of the existing FVD model to address the deficiency of the FVD model in yielding unrealistic and overshooting accelerations/decelerations during certain traffic conditions such as leader lane change where mutation of space headway and approaching rate is likely to happen. All model parameters are also calibrated using field data randomly adopted from the NGSIM program. By doing so, the capped-FVD model is expected to be able to nicely limit down all overshooting accelerations/decelerations yielded by the FVD model so that it can eliminate overreacted vehicular maneuvers and better reproduce following vehicle movements in aforementioned traffic conditions while share the same, good performance as the FVD model in other normal traffic conditions. Performance comparative analyses to validate the performance improvement of the capped-FVD model are included in the other companion paper serving as Part II of this research. At last, the abilities of the capped-FVD model in both preventing rear-end collisions in urgent cases and reproducing traffic flows of stochastic fundamental diagrams [21] will also be investigated as a future work.

References

1. Fang, Z., Li, X., Ma, J.: Parsimonious shooting heuristic for trajectory design of connected automated traffic part I: theoretical analysis with generalized time geography. Transp. Res. Part B Methodol. **95**, S0191261515301806 (2016)
2. Li, X., Ghiasi, A., Xu, A.: A piecewise trajectory optimization model for connected automated vehicles: exact optimization algorithm and queue propagation analysis **118** (2018)
3. Zhou, M., Qu, X., Li, X.: A recurrent neural network based microscopic car following model to predict traffic oscillation. Transp. Res. Part C Emerg. Technol. **84**, 245–264 (2017)
4. Chandler, R.E., Herman, R., Montroll, E.W.: Traffic dynamics: studies in car following. Oper. Res. **6**, 165–184 (1958)
5. Gazis, D.C., Herman, R., Potts, R.B.: Car-following theory of steady-state traffic flow. Oper. Res. **7**, 499–505 (1959)
6. Helly, W.: Simulation of bottlenecks in single-lane traffic flow. In: Theory of Traffic Flow, pp. 207–238 (1959)
7. Gipps, P.G.: A behavioural car-following model for computer simulation. Transp. Res. Part B: Methodol. **15**, 105–111 (1981)
8. Bando, M., Hasebe, K., Nakayama, A., Shibata, A., Sugiyama, Y.: Dynamical model of traffic congestion and numerical simulation. Phys. Rev. E **51**, 1035 (1995)
9. Treiber, M., Hennecke, A., Helbing, D.: Congested traffic states in empirical observations and microscopic simulations. Phys. Rev. E **62**, 1805–1824 (2000)
10. Kesting, A., Treiber, M., Helbing, D.: Enhanced intelligent driver model to access the impact of driving strategies on traffic capacity. Philos. Trans. R. Soc. Lond. A Math. Phys. Eng. Sci. **368**, 4585–4605 (2010)
11. Zhou, M., Qu, X., Jin, S.: On the impact of cooperative autonomous vehicles in improving freeway merging: a modified intelligent driver model-based approach. IEEE Trans. Intell. Transp. Syst. **18**, 1422–1428 (2017)
12. Helbing, D., Tilch, B.: Generalized force model of traffic dynamics. Phys. Rev. E **58**, 133 (1998)
13. Jiang, R., Wu, Q., Zhu, Z.: Full velocity difference model for a car-following theory. Phys. Rev. E **64**, 017101 (2001)
14. Fambro, D., Fitzpatrick, K., Koppa, R.: NCHRP report 400: determination of stopping sight distances. national cooperative highway research program. In: Transportation Research Board, National Research Council. National Academy Press, Washington, DC (1997)
15. W. contributors: Stopping sight distance. https://en.wikipedia.org/wiki/Stopping_sight_distance (2015). Accessed 20 May 2018

16. Framer, D.: Acceleration of a car. https://hypertextbook.com/facts/2001/MeredithBarricella. shtml (2008). Accessed 20 July 2018
17. Kudarauskas, N.: Analysis of emergency braking of a vehicle. Transport **22**, 154–159 (2007)
18. Kesting, A., Treiber, M.: Calibrating car-following models by using trajectory data: methodological study. Transp. Res. Rec. J. Transp. Res. Board 148–156 (2008)
19. FHWA: The next generation simulation (NGSIM). http://www.ngsim.fhwa.dot.gov/ (2008)
20. Thiemann, C., Treiber, M., Kesting, A.: Estimating acceleration and lane-changing dynamics from next generation simulation trajectory data. Transp. Res. Rec. J. Transp. Res. Board 90–101 (2008)
21. Qu, X., Zhang, J., Wang, S.: On the stochastic fundamental diagram for freeway traffic: model development, analytical properties, validation, and extensive applications. Transp. Res. Part B Methodol. **104**, 256–271 (2017)

Chapter 15
Towards Eliminating Overreacted Vehicular Maneuvers: Part II Comparative Analyses

Yang Yu, Yun Zou and Xiaobo Qu

Abstract Microscopic car following models are of great importance to traffic flow studies and vehicular dynamics reproducing. The Full Velocity Difference (FVD) model is a well-known example with satisfactory simulation performances in most times. However, by analyzing the structure of the model formula, we find that it can sometimes generate overreacted vehicular maneuvers such as unrealistically strong (overshooting for short) accelerations or decelerations that conflict with normal driver habits or even beyond the actual vehicular acceleration/deceleration performance, especially when the target vehicle encounter a leader cut-in or move-out (leader lane change for short). As Part II of the entire research, this paper conducts performance comparative analyses between the existing FVD model and the capped Full Velocity Difference (capped-FVD) model introduced in Part I of the research (the other companion paper) to address the above deficiency, and the results indicate that both models are equivalent in most times but the capped-FVD model will outperform the existing FVD model in aforementioned traffic scenarios since overreacted vehicular maneuvers (overshooting accelerations or decelerations) are totally eliminated. In other words, the aforementioned deficiency of the existing FVD model is totally corrected by the capped-FVD model and the capped-FVD model is a better choice for simulating vehicle movements in multi-lane roadways.

Y. Yu (✉) · Y. Zou
School of Civil and Environmental Engineering, University of Technology Sydney, Sydney, NSW 2007, Australia
e-mail: Yang.Yu-4@student.uts.edu.au

Y. Zou
e-mail: Yun.Zou@student.uts.edu.au

X. Qu
Department of Architecture and Civil Engineering, Chalmers University of Technology, 412 96 Gothenburg, Sweden
e-mail: xiaobo@chalmers.se

© Springer Nature Singapore Pte Ltd. 2019
X. Qu et al. (eds.), *Smart Transportation Systems 2019*, Smart Innovation, Systems and Technologies 149, https://doi.org/10.1007/978-981-13-8683-1_15

15.1 Introduction

Microscopic car following models are used to reproduce how vehicles are following one another on real roadways. They are the foundations of microscopic traffic flow theories and is the most important tools in traffic modelling. While some of the latest car following models and algorithms [1, 2] dedicatedly focus on simulating connected automated vehicles, traditional car following models mainly aims at reproducing behaviours of manual vehicles and are normally formula-based that take a vehicle's current status and surrounding traffic conditions as inputs and output the movement of the vehicle at next time step mainly in the form of acceleration rate since acceleration is directly linked to control of engine power [3]. Two early examples include the Gazis–Herman–Rothery (GHR) model [4, 5] and the Linear (Helly) model [6], which were followed by some famous descendants such as the Gipps model [7], the Optimal Velocity (OV) model [8], and the Intelligent Driver Model (IDM) family [9–11]. Among these well-known formula-based car following models, the OV model has the simplest model structure yet can still reproduce vehicular dynamics of normal drivers. The OV model simply considers the impact of the immediate leader and assume that a driver will always identify an optimal velocity based on the space headway (gap) to the immediate leader and attempt to apply different degrees of accelerations according to the velocity difference between the gap-dependent optimal velocity and current actual velocity. The OV model was later improved to a Generalized Force (GF) model by Helbing and Tilch [12] through introducing another term into the OV function which represents the impact of negative velocity difference between the leader and the follower. Traffic simulation result shows that the GF model can less frequently produce too strong accelerations or decelerations but better avoid rear-end collisions than the OV model. Then, the GF model was further upgraded to a Full Velocity Difference (FVD) model by Jiang, et al. [13] to address the unsatisfactory performance of the existing model in reproducing both the delay time of car motion and kinematic wave speed at jam density. Compared with the GF model, the FVD model takes not only the negative leader–follower velocity difference, but also the positive velocity difference into consideration.

Although the FVD model is already a highly improved one in the OV model family, we find that it may produce overshooting accelerations or decelerations that conflict with normal driver habits (Normal drivers will hardly apply accelerations stronger than 3.41 m/s^2 or decelerations stronger than -3.41 m/s^2 [14, 15]) or even beyond the actual vehicular acceleration/deceleration performance (4.50 m/s^2 is the typical maximum acceleration a modern vehicle can reach [16]; -8.41 m/s^2 is the typical maximum deceleration a modern vehicle with anti-lock braking system can reach upon the speed of 60 km/h [17]) in certain traffic scenarios, especially when the target follower encounters leader lane change. In other words, the FVD model may overreact in such traffic scenarios. Thus, the FVD model cannot be applied in multi-lane scenarios where lane changes occur frequently. To address this, we propose a capped-FVD model based on the existing FVD model by introducing an acceleration cap and a deceleration cap, respectively, and limit any overshooting accelerations/decelerations outside the acceleration/deceleration cap to smaller but more reasonable values. By doing so, the capped-FVD model can eliminate overacted vehicular maneuvers and better describe vehicular behaviours in aforementioned traffic scenarios where overshooting accelerations/decelerations may be yielded by the FVD model, and still has the same good performance as the existing FVD model in other normal traffic scenarios.

The rest of the paper (Part II of the research) is organized as follows: Sect. 15.2 briefly introduces the structure of the calibrated capped-FVD model developed in Part I of the research (the other companion paper). Section 15.3 first describes the traffic data adopted for model testing. Then, performance comparative analyses between the capped-FVD model and the existing FVD model are conducted to validate the improvement of the modified model. Section 15.4 shows a conclusion of this paper along with the directions of future works.

15.2   Model Description–The Capped Full Velocity Difference (Capped-FVD) Model

The capped-FVD model is based on the FVD model and it takes the actual velocity of the follower, the gap-dependent optimal velocity of the follower, and the velocity difference between the leader and the follower (approaching rate) as input and yield the acceleration rate of the follower as output. The main function is listed as follows:

$$\ddot{x}_n(t) = \frac{1}{\tau}[V(\Delta x) - \dot{x}_n(t)] + \lambda \Delta v \qquad (15.1)$$

Besides, a general form of the optimal velocity function $V(\Delta x)$ is determined as follows based on the works of Helbing and Tilch [12] and Kesting and Treiber [18]:

$$V(\Delta x) = V_1 + V_2 \tanh\left[\frac{(\Delta x - l)}{l_{int}} - \beta\right] \qquad (15.2)$$

The capped-FVD model also includes another function to restrict any overshooting accelerations or decelerations generated by Eq. (15.1) to a reasonable range, which is displayed as follows:

Table 15.1 Calibration results for the FVD model and the capped-FVD model

	τ	λ	V_1	V_2	l_{int}	β	a_{cap}	d_{cap}	r
FVD model	2.94	0.89	14.41	15.89	15.22	1.51			
Capped-FVD model	2.94	0.89	14.41	15.89	15.22	1.51	3.41	-3.41	3

$$a_n(t) = \begin{cases} a_{cap} : \ddot{x}_n(t) > a_{cap} \cap t = 1 \\ a_{cap} : \ddot{x}_n(t) > a_{cap} \cap \ddot{x}_n(t-1) \le a_{cap} \\ a_{cap} : \ddot{x}_n(t) > a_{cap} \cap \ddot{x}_n(t-1) > a_{cap} \cap \frac{\ddot{x}_n(t)}{\ddot{x}_n^+} > 1 \\ a_{cap} \sqrt[r]{\frac{\ddot{x}_n(t)}{\ddot{x}_n^+}} : \ddot{x}_n(t) > a_{cap} \cap \ddot{x}_n(t-1) > a_{cap} \cap 0 < \frac{\ddot{x}_n(t)}{\ddot{x}_n^+} \le 1 \\ \ddot{x}_n(t) : \ddot{x}_n(t) \subset [d_{cap}, a_{cap}] \\ d_{cap} : \ddot{x}_n(t) < d_{cap} \cap t = 1 \\ d_{cap} : \ddot{x}_n(t) < d_{cap} \cap \ddot{x}_n(t-1) \ge d_{cap} \\ d_{cap} : \ddot{x}_n < d_{cap} \cap \ddot{x}_n(t-1) < d_{cap} \cap \frac{\ddot{x}_n(t)}{\ddot{x}_n^-} > 1 \\ d_{cap} \sqrt[r]{\frac{\ddot{x}_n(t)}{\ddot{x}_n^-}} : \ddot{x}_n(t) < d_{cap} \cap \ddot{x}_n(t-1) < d_{cap} \cap 0 < \frac{\ddot{x}_n(t)}{\ddot{x}_n^-} \le 1 \end{cases} \qquad (15.3)$$

Equations (15.1)–(15.3) compose the capped-FVD model and it is easy to conclude that the modified model still adopts the same logic of the existing FVD model to generate accelerations or decelerations but would further confine any overshooting accelerations or decelerations to a reasonable range $-[d_{cap}, a_{cap}]$ m/s^2. All the model parameters are calibrated and determined in Part I of the research based on field data from the NGSIM program [19] and the calibration results are listed below in Table 15.1.

15.3 Model Performance Comparative Analyses

15.3.1 Data Description and Pre-processing

Same as Part I of the research, we adopt traffic data from NGSIM program to do model testing. We randomly adopt 3 data pairs depicting the movements of 3 different following vehicles and their leaders from either the US Highway 101 (US 101) dataset or the Interstate 80 (I-80) dataset. The data collecting frequency is 0.1 s and each data pair contains the trajectories, velocities and accelerations of both the leader and the follower.

To eliminate mutated and false position values (reduce the impact of measurement errors) and smooth the trajectories, we also apply a symmetric exponential moving average filter [20] with a smoothing width of 0.5 s to all the raw trajectory data. Then, the smoothed velocity/acceleration at each time step are calculated from the first/second order differentiations of the corresponding position values. Besides, all the trajectory data chosen start from 2 s before the lane change moment for the convenience of comparison. Finally, we further limit the smoothed accelerations/decelerations to a range of $[-3.41, 3.41]$ m/s^2, which is consistent with both the acceleration/deceleration range in original NGSIM datasets and the deceleration rate used in determining the stopping sight distance [14, 15].

15.3.2   Performance Comparative Analyses Between Both Models

In this subsection, we use both the calibrated capped-FVD model and the existing FVD model to simulate the trajectories of the 3 different followers in the testing dataset in order to validate the performance improvement of the proposed model. Specifically, both models will first predict the acceleration of the following vehicle and then the simulated trajectory of the follower is acquired by calculating the second order integral of the predicted accelerations (velocity is the first order integral of accelerations). The trajectory simulation accuracy is determined by the errors in the space headway and the mixed error measure [18] is used as the objective function to measure the errors quantitatively. To validate the performance improvement of the proposed model in different traffic scenarios, the three leader–follower data pairs in the testing dataset describe 3 different traffic scenarios, respectively:

1. Follower 1 always follow the same leader, which corresponds to no mutation in space headway Δx and approaching rate Δv;
2. Follower 2 first follows a slow leader and then the leader moves out to adjacent lane and the new leader ahead has higher velocity, which corresponds to mutations (sudden increase) in Δx and Δv;
3. Follower 3 first follows a fast leader and then another vehicle with a smaller velocity cuts in from adjacent lane and become the new leader, which corresponds to mutations (sudden decrease) in Δx and Δv.

The simulated trajectory curves are displayed in Fig. 15.1 while the corresponding simulated acceleration/velocity curves are displayed in Fig. 15.2.

As can be observed from the first graph of Fig. 15.1, the trajectories of follower 1 simulated by both models overlap with each other and they both highly resemble the actual trajectory. The same happens to simulated accelerations/velocities according to the first two graphs in Fig. 15.2. The above indicates that both models are totally equivalent in simulating follower 1. This is because the simulated accelerations of follower 1 yielded by Eq. (15.1) never exceeds the range of $[-3.41, 3.41]$ m/s^2 ($[d_{cap}, a_{cap}]$), as can be seen from the first graph of Fig. 15.2. Therefore, the capped model is reduced to the FVD model according to Eq. (15.3).

However, if we observe the simulated trajectories and corresponding accelerations/velocities of follower 2 and 3, we can find that the capped-FVD model outperforms the FVD model when the target follower encountering leader lane changes. For follower 2, the strongest, impossible-to-reach overshooting acceleration (17.66 m/s^2) yielded by FVD model right after the leader moves out is limited down to a much

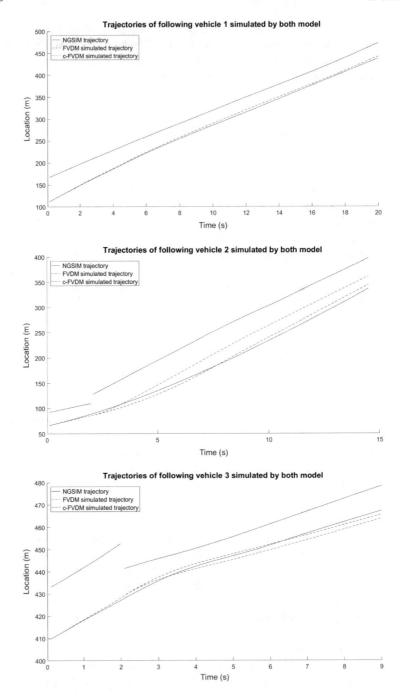

Fig. 15.1 Comparisons of both models in simulating follower trajectories

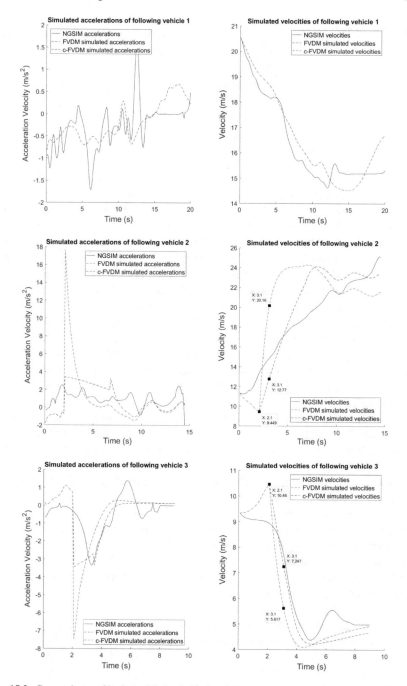

Fig. 15.2 Comparisons of both models in simulating follower accelerations/velocities

Table 15.2 Mixed errors of simulated trajectories by both models

	Follower 1	Follower 2	Follower 3
FVD model mixed error (%)	9.10	34.59	17.86
Capped-FVD model mixed error (%)	9.10	10.28	9.23
Improvement percentage (%)	0	70.28	48.32

more reasonable rate 3.41 m/s^2 (a_{cap}) by the capped-FVD model, and all the subsequent overshooting accelerations are also reduced down to a reasonable but still high rates below 3.41 m/s^2 (a_{cap}) based on their ratio to the strongest overshooting acceleration (17.66 m/s^2). For follower 3, the extremely strong decelerations (up to -7.43m/s^2) yielded by the FVD model when another leader cuts in are also reduced down below -3.41m/s^2 (d_{cap}) by the capped-FVD model according to their ratio to the strongest deceleration (-7.43m/s^2). Therefore, the acceleration (deceleration) curve of follower 2 (follower 3) simulated by the capped-FVD model better matches the corresponding actual curve than that simulated by the FVD model. Since velocity and trajectory is the first and second order integral of acceleration, respectively, the simulated velocity/trajectory curves of both follower 2 and 3 simulated by the capped-FVD model are also much closer to the corresponding actual curves than those simulated by the existing FVD model, as can be observed from Figs. 15.1 and 15.2. The mixed errors of simulated trajectories for all the above followers are listed in Table 15.2.

Please also note that in Fig. 2, the acceleration (deceleration) curve of follower 2 (follower 3) simulated by the capped-FVD model after the leader lane change presents a trend of gradual and slight decrease, which reflect the fact that a relatively strong acceleration or deceleration would be applied by most drivers as the initial response to a sudden change in traffic condition, but it would gradually reduce down for some degrees as the driver gradually adapts to the new traffic condition and thus release the acceleration or braking pedal a bit. When no overshooting accelerations or decelerations are yielded by Eq. (15.1) anymore, which corresponds to the end of special traffic conditions, drivers will be able to adjust the acceleration or deceleration freely again based on the existing FVD model. This is why there is also a small acceleration (deceleration) mutation several seconds after the leader lane change in the simulated acceleration (deceleration) curve of follower 2 (follower 3), as can be observed from Fig. 15.2.

As Table 15.2 shows, both models present the same and very low errors (only 9.1%) in simulating the trajectory of follower 1, which is a strong proof that both models are totally equivalent and very accurate in reproducing following vehicle movements in normal traffic conditions where no overshooting accelerations/decelerations (no mutation of Δx and Δv) are involved. Moreover, the mixed errors of trajectories simulated by the capped-FVD model for both follower 2 and 3 (about 10%) are much lower than those of the existing FVD model (34.59% for follower 2 and 17.86% for follower 3 respectively), and the performance improvement percentages are identified as 70.28 and 48.32%, respectively. This indicates that the capped-FVD model normally has better performances in reproducing following vehicle movements in traffic conditions such as leader lane changes since all unrealistic and overshooting accelerations/decelerations (overreacted vehicular maneuvers) that are yielded by the existing FVD model are nicely eliminated by Eq. (15.3).

At last, it is also worth mentioning that the capped-FVD model does not necessarily outperform the existing FVD model even if the target follower encounters leader lane changes since according to Eq. (15.3), the capped-FVD model will still adopt the same output of the FVD model if no overshooting accelerations/decelerations are yielded by the FVD model, and the mutation of space headway Δx and approaching rate Δv in traffic conditions such as leader lane changes does not necessarily cause such overshooting accelerations/decelerations. In other words, the capped-FVD model will have better performances only if overshooting accelerations/decelerations are involved, regardless of the types of traffic conditions.

15.4 Conclusion

As Part II of this research, performance comparative analyses between the capped-FVD model developed in Part I of the research and the FVD model are conducted in this paper. The results indicate that the capped-FVD model can nicely limit down all overshooting accelerations/decelerations yielded by the FVD model so that it can eliminate overreacted vehicular maneuvers and better reproduce following vehicle movements in multi-lane traffic scenarios while share the same, good performance as the FVD model in other normal traffic conditions. Thus, the capped-FVD model is a better choice for simulating vehicle movements in multi-lane roadways.

Despite the good performance of the capped-FVD model in general, the model has a scope that it may also prohibit the vehicle to yield very strong decelerations in urgent traffic conditions because its overshooting deceleration limitation function (second half of Eq. (15.3)) has set a maximum deceleration rate. This strong but reasonable rate is able to guarantee safety in most cases but may not be sufficient to prevent accidents in urgent cases. Therefore, as a part of the future works, we plan to further modify the model so that it can both eliminate overreacted vehicular maneuvers in non-urgent cases and yield strong enough decelerations to prevent accidents in urgent cases. Moreover, the ability of the capped-FVD model in reproducing traffic flows of stochastic fundamental diagrams [21] needs to be investigated in the future works as well.

References

1. Fang, Z., Li, X., Ma, J.: Parsimonious shooting heuristic for trajectory design of connected automated traffic part I: Theoretical analysis with generalized time geography. Transp. Res. Part B Methodol. **95**, S0191261515301806 (2016)
2. Li, X., Ghiasi, A., Xu, Z.: A piecewise trajectory optimization model for connected automated vehicles: exact optimization algorithm and queue propagation analysis. 118 (2018)
3. Zhou, M., Qu, X., Li, X.: A recurrent neural network based microscopic car following model to predict traffic oscillation. Transp. Res. Part C: Emerg. Technol. **84**, 245–264 (2017)
4. Chandler, R.E., Herman, R., Montroll, E.W.: Traffic dynamics: studies in car following. Oper. Res. **6**, 165–184 (1958)
5. Gazis, D.C., Herman, R., Potts, R.B.: Car-following theory of steady-state traffic flow. Oper. Res. **7**, 499–505 (1959)
6. Helly, W.: Simulation of bottlenecks in single-lane traffic flow. Theory Traffic Flow, 207–238 (1959)
7. Gipps, P.G.: A behavioural car-following model for computer simulation. Transp. Res. Part B: Methodol. **15**, 105–111 (1981)
8. Bando, M., Hasebe, K., Nakayama, A., Shibata, A., Sugiyama, Y.: Dynamical model of traffic congestion and numerical simulation. Phys. Rev. E **51**, 1035 (1995)
9. Treiber, M., Hennecke, A., Helbing, D.: Congested traffic states in empirical observations and microscopic simulations. Phys. Rev. E **62**, 1805–1824 (2000)
10. Kesting, A., Treiber, M., Helbing, D.: Enhanced intelligent driver model to access the impact of driving strategies on traffic capacity. Philos. Trans. R. Soc. Lond. A: Math., Phys. Eng. Sci. **368**, 4585–4605 (2010)
11. Zhou, M., Qu, X., Jin, S.: On the impact of cooperative autonomous vehicles in improving freeway merging: a modified intelligent driver model-based approach. IEEE Trans. Intell. Transp. Syst. **18**, 1422–1428 (2017)
12. Helbing, D., Tilch, B.: Generalized force model of traffic dynamics. Phys. Rev. E **58**, 133 (1998)
13. Jiang, R., Wu, Q., Zhu, Z.: Full velocity difference model for a car-following theory. Phys. Rev. E **64**, 017101 (2001)
14. Fambro, D., Fitzpatrick, K., Koppa, R.: NCHRP Report 400: Determination of stopping sight distances. national cooperative highway research program. In: Transportation Research Board, National Research Council. National Academy Press, Washington (1997)
15. Contributors, W.: Stopping sight distance (2015, 20 May 2018). https://en.wikipedia.org/wiki/Stopping_sight_distance
16. Framer, D.: Acceleration of a Car (2008, 20 May 2018). https://hypertextbook.com/facts/2001/MeredithBarricella.shtml
17. Kudarauskas, N.: Analysis of emergency braking of a vehicle. Transport **22**, 154–159 (2007)
18. Kesting, A., Treiber, M.: Calibrating car-following models by using trajectory data: Methodological study. Transp. Res. Rec.: J. Transp. Res. Board 148–156 (2008)
19. FHWA. The Next Generation Simulation (NGSIM) (2008). http://www.ngsim.fhwa.dot.gov/
20. Thiemann, C., Treiber, M., Kesting, A.: Estimating acceleration and lane-changing dynamics from next generation simulation trajectory data. Transp. Res. Rec.: J. Transp. Res. Board, 90–101 (2008)
21. Qu, X., Zhang, J., Wang, S.: On the stochastic fundamental diagram for freeway traffic: model development, analytical properties, validation, and extensive applications. Transp. Res. Part B: Methodol. **104**, 256–271 (2017)

Chapter 16
Unsupervised Deep Learning to Explore Streetscape Factors Associated with Urban Cyclist Safety

Haifeng Zhao, Jasper S. Wijnands, Kerry A. Nice, Jason Thompson, Gideon D. P. A. Aschwanden, Mark Stevenson and Jingqiu Guo

Abstract Cycling is associated with health, environmental and societal benefits. Urban infrastructure design catering to cyclists' safety can potentially reduce cyclist crashes and therefore, injury and/or mortality. This research uses publicly available big data such as maps and satellite images to capture information of the environment of cyclist crashes. Deep learning methods, such as generative adversarial networks (GANs), learn from these datasets and explore factors associated with cyclist crashes. This assumes existing environmental patterns for roads at locations with and without cyclist crashes, and suggests a deep learning method is able to learn the hidden features from map and satellite images and model the road environments using GANs. Experiments validated the method by identifying factors associated with cyclist crashes that show agreement with existing literature. Additionally, it

H. Zhao (✉) · J. S. Wijnands · K. A. Nice · J. Thompson · G. D. P. A. Aschwanden ·
M. Stevenson
Transport, Health and Urban Design Research Hub, Melbourne School of Design, The University of Melbourne, Parkville, VIC 3010, Australia
e-mail: haifeng.zhao@unimelb.edu.au

J. S. Wijnands
e-mail: jasper.wijnands@unimelb.edu.au

K. A. Nice
e-mail: kerry.nice@unimelb.edu.au

J. Thompson
e-mail: jason.thompson@unimelb.edu.au

G. D. P. A. Aschwanden
e-mail: gideon.aschwanden@unimelb.edu.au

M. Stevenson
e-mail: mark.stevenson@unimelb.edu.au

J. Guo
Key Laboratory of Road and Traffic Engineering of the Ministry of Education, Tongji University, Shanghai 200092, China
e-mail: guojingqiu@hotmail.com

© Springer Nature Singapore Pte Ltd. 2019
X. Qu et al. (eds.), *Smart Transportation Systems 2019*, Smart Innovation, Systems and Technologies 149, https://doi.org/10.1007/978-981-13-8683-1_16

revealed the potential of this method to identify implicit factors that have not been previously identified in the existing literature. These results provide visual indications about what streetscapes are safer for cyclist and suggestions on how city streetscapes should be planned or reconstructed to improve it.

16.1 Introduction

Cycling as an active and sustainable transportation mode is associated with health, environmental and societal benefits. Increased uptake of cycling reduces congestion and pollution, decreases energy consumption and supports healthy and sustainable lifestyles [1]. Increasing the use of bicycles is being supported as a transport policy in many countries [2]. However, cyclists are vulnerable road users and are over-represented in crash statistics [3]. The crash risk associated with cycling discourages people from adopting cycling as a main transportation mode [4] and improving both perceived and actual cycling safety, is key to promoting cycling [5].

Considerable research related to cycling safety focuses on intersections [6], round-abouts [7], and cycle tracks [8] using traditional methods applying either in-person surveys or video footage observation and analysis; all very time consuming and expensive approaches [9]. Until now machine learning, in particular, the use of deep neural networks, has seldom been applied to the study of cyclist crashes.

While particular reasons may apply to a specific cyclist crash, the aggregation of crashes across time and multiple locations infers an array of relations between the road environment and cyclist crashes. We hypothesise that there exist different high-level hidden features in road environments of crash domain and non-crash domain that are difficult to identify by examining individual crash locations. To identify these hidden patterns, we use GANs, a class of algorithms for unsupervised machine learning, that are able to model the distribution of unstructured datasets with the potential to extract high-level features. GANs were first developed by [10] to reconstruct images, and have been widely applied to convert low-resolution images into high-resolution images [11], inpainting [12] and style transfer [13]. In particular, Wijnands et al. [14] suggested GANs as a method for computer-generated design interventions to improve citizens' health and well-being.

Publicly available big data such as street view images, maps and satellite images capture road environments from different perspectives and provide information of different abstraction. Street view images offer street-level imagery and give detailed information about objects in the streetscape. On the other hand, maps are more abstract and semantically rich, and provide not only locations and boundaries of visible objects such as roads, parks, buildings, rivers and facilities, but can also provide additional information such as public transport routes and terrain. Finally, satellite images capture widespread heterogeneous information that is usually not captured by maps, such as lane markings, road surface types, the density of trees and the shadow of buildings. In this paper, we use GANs to encode the road environments contained in map images, and to extract the hidden features for urban road environments of crash and non-crash domains. The results provide unique insights between road environments of crash and non-crash domains.

16.2   Methodology

16.2.1   Data Source

The study region is the Greater Melbourne metropolitan area (Australia), which has a population of almost five million. Of special note to this study, Melbourne has the largest urban tram network in the world [15].

This paper uses publicly available datasets for cyclist crashes recorded by the state road authority—VicRoads. In total, 5,156 crashes were recorded from January 2010 to December 2013 in Greater Melbourne (Fig. 16.1a). For exposure data, we used the anonymized bicycle trips recorded by volunteer users of the RiderLog smartphone application from 2010 to 2013 [16].

Google Maps and satellite images, were downloaded at sampled locations at a resolution of 320 × 320 using the Google Maps Static API. We used a zoom level of 19 and turned off the visibility of all labels. For Google Maps images, we eliminated most geographic characteristics and showed only parks, roads, transit lines, transit stations and bodies of water. We defined a colour scheme for Google Maps images

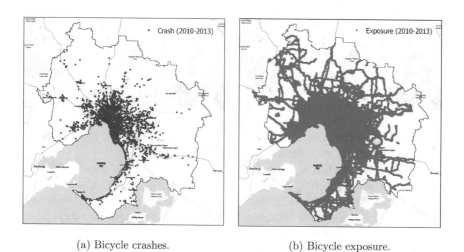

(a) Bicycle crashes.   (b) Bicycle exposure.

Fig. 16.1   Cyclist crashes and bicycle exposure in Greater Melbourne

Fig. 16.2 Example of a
Google Maps image

of white for image background, black for roads, blue for water bodies, orange for
transit lines and green for parks (e.g. see Fig. 16.2).

16.2.2 GANs

GANs are neural network architectures, which simultaneously train a generative
network called generator and a discriminative network called discriminator. The
generator is a decoder function, but instead of reconstructing the inputs, it can work
on any representation in the latent space. The discriminator takes both the generated
images and images from the training dataset as inputs, and returns probabilities
indicating how likely the input images are true images.

The generator's training objective is to generate images that cannot be identified
as fake by the discriminator, while the discriminator tries to improve its capacity
of detecting the fake images. During training, the generator learns to model the
data distribution of images from the training dataset, and the discriminator learns to
model the boundary of data distributions between images from the training dataset
and images produced by the generator.

For a well-trained GAN, if the input is an image from one domain, then the
generator can be used not only to reconstruct the input images back to its own domain,
but also to generate images into a different domain (i.e. image translation). To observe
the differences between crash domain and non-crash domain, we can translate images
from one domain to the other and compare the changes of those images before and
after translations. This paper used the GAN implementation developed by Liu et al.
[17].

16.3   Experiment

In this experiment, we sampled training datasets from only on-road bicycle networks. In total, 1,712 locations were sampled for crash domain, and 2,061 locations were sampled for non-crash domain.

Translation from non-crash domain to crash domain identified a few factors: Tram lines are associated with more cyclist crashes (Fig. 16.3a, b); green spaces are associated with a decrease in cyclist crashes (Fig. 16.3a, c); and road environments with high-rise buildings on roadsides casting big shadows are associated with increased crashes (Fig. 16.3d created a high-rise building using concrete ground for car parking). Figure 16.3e removed the median strip that separates traffic from opposing lanes on divided roadways and placed pavements at the same position. Similarly, in

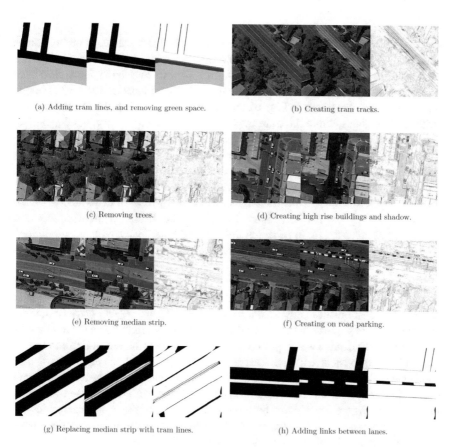

(a) Adding tram lines, and removing green space.   (b) Creating tram tracks.

(c) Removing trees.   (d) Creating high rise buildings and shadow.

(e) Removing median strip.   (f) Creating on road parking.

(g) Replacing median strip with tram lines.   (h) Adding links between lanes.

Fig. 16.3 Themes observed when translating images from non-crash to crash domain. Each subfigure shows three images: the input image (left), the generated image (middle) and (right) the differences between the input image and the generated image

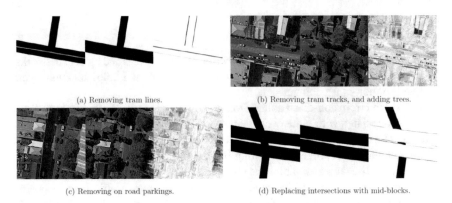

(a) Removing tram lines.   (b) Removing tram tracks, and adding trees.

(c) Removing on road parkings.   (d) Replacing intersections with mid-blocks.

Fig. 16.4 As Fig. 16.3, but from crash domain to non-crash domain

Fig. 16.3f, the median strips that separate tram tracks from roadways on both sides were removed, and the median strip on one side was replaced by on-road parking.

Another key observation is that the tram tracks in Fig. 16.3f have also been removed and replaced by roadways. Note that, the tram tracks in Fig. 16.3b are different from the tram tracks in Fig. 16.3f. The tram tracks in Fig. 16.3f have median strips on both sides, while the tram tracks in Fig. 16.3b do not have median strips. The tram tracks in Fig. 16.3b share road space with vehicles, i.e. vehicles can directly run on tram tracks. Comparing the translation in Fig. 16.3b which created tram tracks on existing roadways and the translation in Fig. 16.3f which removed tram tracks, we identify that tram tracks which share road space with vehicles are associated with increased cyclist crashes, while tram tracks separated by median strips are associated with decreased cyclist crashes. Figure 16.3g in which median strips were replaced by tram lines and Fig. 16.3h in which links were created between lanes on divided roadways both confirm that separated traffic is associated with decreased cyclist crashes.

The reverse translation from crash domain to non-crash domain confirms that the removal of tram tracks which share road space with vehicles are associated with decreased cyclist crashes (Fig. 16.4a, b). From Fig. 16.4b, we can observe that the tram tracks share space with vehicles, and two vehicles are running on the tram tracks when the satellite image was captured. Figure 16.4c shows the removal of road parking, which is consistent with the reverse translation in Fig. 16.3f. Figure 16.4c also shows the replacement of high-rise buildings with low buildings in the translation and the shadow with trees on-road sides, which is consistent with the reverse translation in Fig. 16.3d. Figure 16.4b and c both show an increase of green space. Besides the above themes which are consistent with the reverse translation, intersections have been replaced by mid-blocks in some translations (Fig. 16.4d).

Figure 16.4b and c also show a change of roof colour of buildings from silver to brick red. In Melbourne, most of the buildings in business or industry areas are high-rise buildings with silver colour top, while most of the residential areas have lower buildings with brick red colour. Residential areas generally have low traffic than business areas. Traffic factors have not been considered in this experiment, but the availability of high-quality exposure data and traffic volumes data will allow the method to explore those factors in more specific areas such as highly used intersections in CBD areas, and produce more detailed results.

16.4   Discussions

16.4.1   Summary

This experiment sampled training datasets from on-road bicycle networks. For on-road bicycle networks:

- Tram tracks are generally associated with more cyclist crashes, except for tram tracks that are separated from other traffic.
- Road environments for non-crash locations tend to have more green space (trees or grass), may indicative of density.
- Road environments with high-rise buildings casting shadows on roadsides are mostly seen in the crash domain.
- Intersections are associated with more cyclist crashes.
- On-road parking is associated with more cyclist crashes.
- Median strips are associated with reduced cyclist crashes.

Generally, these identified factors have been suggested previously in other studies, providing confidence in the presented methodology. Vandenbulcke et al. [9] found that a high risk is statistically associated with the presence of on-road tram tracks, complex intersections, and proximity to shopping centres or garages. Several other studies have also confirmed that tram tracks are strongly associated with bicycle crashes [15]. For example, Teschke et al. [18] found that for cities with extensive use of trams (or streetcars), one-third of cyclist crashes had directly involved tram or train tracks. Melbourne has the largest mixed traffic tram network in the world. It has also been found that on-street parking is associated with increased crash risk for on-road cycling [2], and separating bicycles and motor vehicles prevent the two modes from colliding [3].

16.4.2   Interpretation

The method is not intended to provide direct solutions to the problem of cyclist crashes, and does not diagnose the cause of cyclist crashes, which can be very diverse and are specified on the accident report from police, but investigates underlying factors. This method is an analytic tool that takes advantage of the increasing availability of big datasets, computing power and the advances of deep learning techniques, to analyse the road environments or even neighbourhood of locations that have cyclist crashes from a new perspective.

Generally, this method helps to achieve the fundamental target of still reducing cyclist crashes while promoting more cycling by serving three purposes:

- The method helps gain a better understanding of what factors are associated with cyclist crashes.
- The method can provide suggestions on what road environments are more likely to have less cyclist crashes than others, therefore redirecting cyclists to take certain routes.
- The method can provide suggestions on how to design city streets to reduce cyclist crashes.

16.4.3   Limitations

This paper investigates the factors associated with bicycle crashes based on imagery training, so the factors that can be identified are limited by the type of images. For example, training Google Maps images cannot give information about whether buildings are factors associated with cyclist crashes, because buildings are not shown in Google Maps images in the training datasets. This paper identified factors related to infrastructure characteristics. Cyclists characteristics such as gender and age [19] that are not in the images cannot be identified by this method.

Another limitation of this method is the incompleteness of the dataset, especially the crash data and exposure data. The officially reported number of crashes can be significantly lower than the actual number of crashes [20]. The exposure data does not cover all the trips. The incomplete record of crashes and exposure provides a generic indication upon where the crashes and trips are more and where are less, therefore the data can still be used for training purposes. There is still potential that the incompleteness of the records leads to biased training results. A complete record of the crashes and exposure will lead to more accurate outcomes. However, high-quality exposure data is difficult to obtain [21]. Appropriate exposure measurements and crash measurements are fundamental questions to be addressed [22].

16.5   Conclusion

This paper presents a deep learning method to model road environments and identify factors that are associated with cycling crashes. The factors include tram tracks, on-street parking, intersections, on-road bicycle lanes, median strips and green spaces. The method has been validated by an experiment that identifies factors as anticipated. The results confirm that factors associated with bicycle crashes can be identified by modelling and comparing streetscape imagery for crash and non-crash domains using GANs. This paper contributes to the literature by providing an analytic tool to assess the road environments and cyclist crashes. The method provides insights to urban designers, infrastructure planners and cyclist towards how to improve cycling safety and prevent crashes. Since images data such as Google Maps and satellite images are available worldwide, the method can be readily applied to any other country and region.

The method identifies factors associated with cyclist crashes by analysing imagery of urban areas. The advantage of this method is that it is not limited to one or two factors, rather it explores all potential factors that may be associated with bicycle crashes. The main limit to the range of factors it can explore is the amount of information the images provided. The method is purely based on objective imagery information; therefore, it eliminates the possible bias in traditional human participated research methods. The method is not limited to bicycle safety, and can readily be applied to other research domains such as pedestrian safety and vehicle safety. Due to the limited information provided by satellite images and Google Maps, the factors identified in this paper are mainly related to the urban infrastructure and urban environment.

Exposure and crash data are the key factors that may affect the training results. The exposure data may not cover all real trips, and the number of crashes is often under-reported. The availability of a complete dataset will improve the accuracy of the results. Future work can also incorporate other data sources such as vehicle volumes, and train images from other sources such as Google Street View images or maps containing traffic information. Future research may investigate other modes such as pedestrian safety and vehicle safety, and focus on specific types of infrastructure only such as intersections and roundabouts.

References

1. Cheng, W., Gill, G.S., Ensch, J.L., Kwong, J., Jia, X.: Multimodal crash frequency modeling: multivariate space-time models with alternate spatiotemporal interactions. Accid. Anal. Prev. **113**, 159–170 (2018)
2. Beck, B., Stevenson, M., Newstead, S., Cameron, P., Judson, R., Edwards, E.R., Bucknill, A., Johnson, M., Gabbe, B.: Bicycling crash char- acteristics: an in-depth crash investigation study. Accid. Anal. Prev. **96**, 219–227 (2016)
3. DiGioia, J., Watkins, K.E., Xu, Y., Rodgers, M., Guensler, R.: Safety impacts of bicycle infrastructure: a critical review. J. Saf. Res. **61**, 105–119 (2017)
4. Prato, C.G., Kaplan, S., Rasmussen, T.K., Hels, T.: Infrastructure and spatial effects on the frequency of cyclist-motorist collisions in the Copenhagen Region. J. Transp. Saf. Secur. **8**, 346–360 (2016)
5. Silla, A., Leden, L., R¨am¨a, P., Scholliers, J., Van Noort, M., Bell, D.: Can cyclist safety be improved with intelligent transport systems? Accid. Anal. Prev. **105**, 134–145 (2017)
6. Asgarzadeh, M., Verma, S., Mekary, R.A., Courtney, T.K., Christiani, D.C.: The role of intersection and street design on severity of bicycle-motor vehicle crashes. Inj. Prev. **23**, 179–185 (2017)

7. De Rome, L., Boufous, S., Georgeson, T., Senserrick, T., Richardson, D., Ivers, R.: Bicycle crashes in different riding environments in the Australian capital territory. Traffic Inj. Prev. **15**, 81–88. https://doi.org/10.1080/15389588.2013.781591

8. Thomas, B., DeRobertis, M., Board, T.R.: The safety of urban cycle tracks: a review of the literature. Accid. Anal. Prev. **52**, 6p (2012)

9. Vandenbulcke, G., Thomas, I., Int Panis, L.: Predicting cycling acci- dent risk in Brussels: a spatial case-control approach. Accid. Anal. Prev. **62**, 341–357 (2014)

10. Goodfellow, I.J., Pouget-Abadie, J., Mirza, M., Xu, B., Warde-Farley, D., Ozair, S., Courville, A., Bengio, Y.: Generative Adversarial Networks, pp. 1–9 (2014)

11. Sønderby, C.K., Caballero, J., Theis, L., Shi, W., Husz´ar, F.: Amortised map inference for image super-resolution. arXiv:1610.04490 (2016)

12. Pathak, D., Krahenbuhl, P., Donahue, J., Darrell, T., Efros, A.A.: Context encoders: feature learning by inpainting. In: Proceedings of the IEEE Conference on Computer Vision and Pattern Recognition, pp. 2536–2544 (2016)

13. Gatys, L.A., Ecker, A.S., Bethge, M.: Image style transfer using convolutional neural networks. In: Proceedings of the IEEE Conference on Computer Vision and Pattern Recognition, pp. 2414–2423 (2016)

14. Wijnands, J., Nice, K., Thompson, J., Zhao, H., Stevenson, M.: Streetscape augmentation using generative adversarial networks: insights related to health and wellbeing. arXiv:1905.06464 (2019)

15. Naznin, F., Currie, G., Logan, D.: Exploring the key challenges in tram driving and crash risk factors on the Melbourne tram network: tram driver focus groups. Australas. Transp. Res. Forum **16**, 1–15 (2016)

16. Leao, S.Z., Pettit, C.J.: RiderLog Anonymised Bicycling Data (2017)

17. Liu, M., Breuel, T., Kautz, J.: Unsupervised image-to-image transla- tion networks. arXiv: 1703.00848 (2017)

18. Teschke, K., Dennis, J., Reynolds, C.C., Winters, M., Harris, M.A.: Bicycling crashes on streetcar (tram) or train tracks: mixed methods to identify prevention measures. BMC Public Health **16**, 1–10 (2016)

19. Prati, G., Pietrantoni, L., Fraboni, F.: Using data mining techniques to predict the severity of bicycle crashes. Accid. Anal. Prev. **101**, 44–54 (2017)

20. Juhra, C., Wiesk¨otter, B., Chu, K., Trost, L., Weiss, U., Messerschmidt, M., Malczyk, A., Heckwolf, M., Raschke, M.: Bicycle accidents - do we only see the tip of the iceberg?: A prospective multi-centre study in a large German city combining medical and police data. Injury **43**, 2026–2034 (2012)

21. Vanparijs, J., Int, L., Meeusen, R., De Geus, B.: Exposure measure- ment in bicycle safety analysis: a review of the literature. Accid. Anal. Prev. **84**, 9–19 (2015)

22. Saha, D., Alluri, P., Gan, A., Wu, W.: Spatial analysis of macro-level bicycle crashes using the class of conditional autoregressive models. Accid. Anal. Prev. 0–1 (2018). https://doi.org/10.1016/j.aap.2018.02.014

Chapter 17
On the Impact of Emergency Incidents on the Freeway: A Full Velocity Difference (FVD) Model Based Four-Lane Traffic Dynamics Simulation

Yun Zou and Xiaobo Qu

Abstract Freeway as a relatively close traffic system is extremely vulnerable to bottlenecks, especially those created by emergency incidents, such as traffic accident and vehicle breakdown. Usually, at least one lane will be either partially impacted or fully blocked. To have an in-depth understanding of the impact, we develop a four-lane traffic dynamic model and microscopically demonstrate the impact brought by bottlenecks on the freeway. We also incorporate a realistic FVD-based lane-changing trajectory by approximating the clothoid into cubic polynomials. Then, the lane-change cooperation can be realistically demonstrated as the time span of the lane change is not simplified as one simulation interval. By comparing the dynamic change of trajectories, we found that the communication with the preceding and the following vehicle of target lane are equally important to overall traffic performance.

17.1 Introduction

Transportation has been playing an extremely significant role in modern societies form people's daily routines to the economy of countries. A smooth traffic can bring high satisfaction to all participations in our society, while a traffic jam can significantly impact the functionality of a city. Unfortunately, the truth is that our traffic systems are usually unstable with the presence of incidents, such as traffic accidents or vehicle breakdowns. In that case, lanes are likely to be either fully blocked or partially impacted creating bottlenecks to the surrounding traffic system. Then, vehicles on the blocked lane will be forced to execute lane-changing activities, otherwise, they will stop at the merging point, which inversely influences the lane-changing manoeuvre. It is found that both lane closures and lane-changing activities impact the traffic capacity significantly [8, 11]. With the decreasing of traffic capacity, road

Y. Zou
University of Technology Sydney, 123 Broadway, Ultimo, NSW 2007, Australia
e-mail: yun.zou@student.uts.edu.au

X. Qu (✉)
Chalmers University of Technology, 412 96 Gothenburg, Sweden
e-mail: drxiaoboqu@gmail.com

© Springer Nature Singapore Pte Ltd. 2019
X. Qu et al. (eds.), *Smart Transportation Systems 2019*, Smart Innovation,
Systems and Technologies 149, https://doi.org/10.1007/978-981-13-8683-1_17

users may experience longer travel time, higher speed variation and traffic congestion [16]. Freeways is a specific scenario where the number of entry and exit is relatively limited compared to other traffic systems, are extremely vulnerable to aforementioned bottlenecks. Once an incident occurs on the freeway, drivers cannot avoid the bottleneck via reconfiguring the route, alternatively, they have to queue upstream of the incident area until incident being cleared. Apart from wasting a tremendous amount of time on queuing, there is also a high possibility that drivers experience the stop-and-go traffic condition and poor collaborations from other drivers.

To have an in-depth understanding of the impact brought by incidents, researcher has done a large quantity of simulations both macroscopically and microscopically. Among these researches, microscopic simulation models demonstrate a better performance over macroscopic simulation models with regards to reproducing the cooperation among vehicles, furthermore, the microscopic simulation model can demonstrate how vehicles are impacted by lane-changing motion individually and detailly [15]. Car-following model and lane-change model as two essential components of microscopic simulation model is wildly discussed. FVD model as one of the most prevalent car-following models has been applied and modified to simulate the driving behaviour under single-lane traffic [13]. The development of FVD model can date back to 1995 when Bando et al. [1] proposed the Optimal Velocity (OV) model. In the OV model, an optimal following velocity is calculated for every vehicle by substitute the following gap to their proceeding vehicle into:

$$\ddot{x}_n(t) = \kappa[V(s) - \dot{x}_n(t)] \tag{17.1}$$

where κ denotes the sensitivity constant, $V(s)$ denotes the optimal velocity under the current following gap, $\dot{x}_n(t)$ denotes the current velocity of vehicle n, $\ddot{x}_n(t)$ denotes the acceleration rate during the following simulation interval. It was then calibrated by Helbing and Tilch [3] based on field data, and a function was developed as

$$V(s) = V_1 + V_2 \tanh[C_1(\Delta x - l) - C_2] \tag{17.2}$$

However, it was soon found that OV model can result in unrealistically high acceleration rates and deceleration rates [6]. To limit the acceleration rate and deceleration rate into a realistic range, Jiang [6] proposed the FVD model taking the velocity difference of two successive vehicles into account. However, Yu et al. [14] found that large acceleration and deceleration rates are still observable, thus, he proposed a confined FVD model and calibrated the optimal velocity function based on field data collected from NGSIM. The calibrated values are demonstrated in Table 17.1.

Table 17.1 Calibrated parameter of confined FVD

Parameters	V_1	V_2	C_1	C_2
Values	14.300	15.997	0.066	1.508

Due to the similarity of researched area (freeway) between Yu's research and ours, we adopted his calibrated parameter to simulate the car-following motion of single-lane traffic, and the function of adopted FVD model is

$$\ddot{x}_n(t) = \kappa[V(s) - \dot{x}_n(t)] + \lambda \Delta \dot{x} \qquad (17.3)$$

where $\Delta \dot{x}$ denotes the velocity difference between two successive vehicles, λ denotes the sensitivity as

$$\lambda = \begin{cases} 0: & \Delta x \le critical\ gap \\ 0.5: & \Delta x > critical\ gap \end{cases} \qquad (17.4)$$

Although there is a large quantity of research on car-following model, researchers pay much less attention on lane-changing model at same time, fortunately, lane-changing model has been attracting more and more attention recently, because researchers have been gradually noticing that (1) drivers are prone to make mistake during lane-changing motions and (2) lane-changing motions bring native impact on capacity, reducing the bottlenecks discharge rate [15]. Based on the field observations, Hidas [5] group the lane-changing manoeuvre into free lane change, forced lane change and cooperative lane change, and he discussed the difference among these three types of lane-changing manoeuvres and illustrated the procedures of each lane-changing manoeuvre in detail. Hidas [5] also proposed a lane-changing model comparing the space gap between the preceding vehicle and the following vehicle respectively with the minimum required space gap, however, Hidas [5] made an assumption that lane-changing manoeuvres are executed instantaneously, thus the time span of the lane-changing motion is taken as 1 s which is equal to the simulation interval. This assumption is not always realistic, especially under complex lane-changing situation when subject vehicles travel much more slowly than vehicles in the target lane so that the drivers cannot make the lane-changing decision immediately and determinedly. It is obvious that this model did not discuss the trajectories when vehicles execute lane-changing task, thus, there is a need to develop a lane-changing model taking lane-changing trajectories into account.

In our research, we develop a four-lane microscopic model simulating the traffic dynamics on a four-lane freeway with an incident blocking one lane, as shown in Fig. 17.1.

17.2 Model Development

17.2.1 FVD Model

FVD model as one of the most prevailing car-following models has been widely researched, and it has been proved that FVD model is able to reproduce the car-

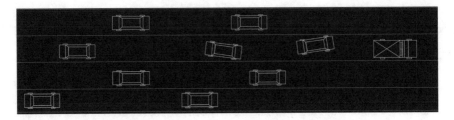

Fig. 17.1 Demonstration of researched four-lane freeway

following motion both microscopically and macroscopically [6, 14]. Hence, we adopt FVD model as the car-following model as illustrated in Eqs. (17.1)–(17.4), and we introduce the calibrated parameters from Yu et al. [14] due to the similarity between the two researched cases. In order to simulate the deceleration motion when vehicle approaching the bottleneck, we propose a quadratic equation to demonstrate the relationship between remaining distance to incident and the velocity. Because the existence of the bottleneck is the main reason for the deceleration motion, we assume that the velocity limit is determined by the remaining distance to the bottleneck impacted by an incident. We, therefore, introduce a velocity cap to limit the maximum velocity of an approaching vehicle as

$$\text{VC}(x(t)) = a_1(x(t) - x_{start})^2 + a_2(x(t) - x_{start}) + a_3 \quad (17.5)$$

where VC(s) denotes the velocity cap, $x(t)$ denotes the longitudinal location at time t, x_{start} denotes the longitudinal location where drivers start to decelerate, a_1, a_2 and a_3 are constants. Before calibrating these parameters, we must clarify that where vehicles will start to be affected by the incident. We, therefore, assume that the distance is consist of the length of advanced warning area L_{aw} and the visibility distance L_{vd}. As L_{aw} and L_{vd} are highly dependent on road geometry, we assumed that approaching vehicles start to decelerate when they are 300 m away from the incident area, nevertheless, the closer the vehicles get the higher the deceleration rates are supposed to be. VC much fulfil the requirement of

$$\begin{cases} VC(x_{start}) = 16.7 \\ VC(x_{incident}) = 0 \end{cases} and \begin{cases} \dot{V}C(x_{start}) \approx 0 \\ \dot{V}C(x_{incident}) \geq -D_{limit} \end{cases} \quad (17.6)$$

where D_{limit} denotes the maximum deceleration, which is assumed to be 2.7 m/s^2 [5], $x_{incident}$ denotes the longitudinal location of incident area. Thus, Eq. (17.5) is calibrated as

$$\text{VC}(x(t)) = -1.86 \times 10^{-4}(x(t) - x_{start})^2 + 16.7 \quad (17.7)$$

17.2.2   Lane-Changing Model

Lane-changing model as an essential component of traffic dynamics simulation model reproduces the interaction among vehicles at complex traffic conditions [5]. Lv et al. [9] proposed a lane-changing model based on FVD car-following model, but the model is oversimplified for two reasons: (1) the critical acceptable gaps are constantly determined by safety distance; (2) the lane changing is finished instantaneously. In Lv's model, drivers are encouraged to execute lane-changing manoeuvres either by higher velocity of larger front gap, while assumption is made in many other bottleneck-related researches that vehicles on the through lane do not merge into the blocked lane, and the only stimulation for lane change is the distance to the bottleneck [4, 10, 16]. Zheng [15] classified the impact brought by lane-changing motion into three groups by their occurrence period which are anticipation, insertion and relaxation. As we assume that driver will commence merging into the target lane without any hesitation as long as the immediate preceding gap and following gap are acceptable, drivers on target lane are unlikely to be disturbed during the anticipation period. However, we do include the reaction time in our lane-changing model by adding a term $C_r V_{platoon}$, which is the distance that the platoon in the target lane moved during the reaction period as shown in

$$\begin{cases} g_{p,min} = a_p l + C_r V_{platoon} \\ g_{f,min} = a_f l + C_a V_{platoon} \end{cases} \tag{17.8}$$

where $g_{p,min}$ and $g_{f,min}$ denote the minimum acceptable preceding gap and following gap, respectively, l denotes the length of vehicles, C_r denotes the reaction factor which is assumed to be 0.25 s, C_a denotes the coefficient related to the acceptable gap and a_f are constant which are assumed to be 1 and 2, respectively (in that case, the safety distance is equal to the length of vehicles). As long as the preceding gap and following gap is greater than the respective minimum acceptable gap, drivers commence lane change determinedly.

17.2.3   Lane-Changing Trajectories

As many previous models are being established under the assumption that lane-changing motion is finished within one simulation interval, they inevitably ignore the lane-changing trajectories [4, 9, 10, 15, 16]. Sledge and Marshek [12] innovatively introduced the concept of emergency lane-changing trajectories, and Slege and Marshek [12] proposed that an ideal lane-changing trajectory need to be continuous and smooth with optimal curvature and minimum longitudinal span. Gackstatter et al. [2] approximated the lane-changing trajectories into polynomials. Kasper et al. [7] introduced a lane-change model with third-order polynomial trajectories as

$$y = A_3(x(t) - x_{start})^3 + A_2(x(t) - x_{start})^2 + A_1(x(t) - x_{start}) + A_0. \quad (17.9)$$

where $A_3 = \frac{-2(y_{end} - y_{start})}{D^3}$, $A_2 = \frac{3(y_{end} - y_{start})}{D^2}$, $A_1 = 0$, $A_0 = y_{start}$, y_{start} and y_{end} denote the lateral position at the start and the end of the lane change respectively, x_{start} denotes the longitudinal location when vehicle commences lane change, D denotes the longitudinal distance of the lane-changing trajectory, and it is restricted by

$$D \geq \dot{x}(t)\sqrt{\frac{6|y_{end} - y_{start}|}{|\ddot{y}_{max}|}} \quad (17.10)$$

where $\ddot{y}_{max}$ denotes the maximum lateral acceleration rate. In order to demonstrate a smooth trajectory, the value of $\dot{x}(t)$ is fixed as the velocity when vehicle starts to change lane.

17.3   Discussion

In the simulation, we hypothesized an incident occurs on lane 2, and the incident is located at the point of 450 m. Vehicles on the blocked lane start to be impacted by the existence of incident from the point of 100 m. It is also assumed that vehicle on through lane will not execute lane-changing motion due to a lack of necessity. Figures 17.2, 17.3, 17.4, 17.5 and 17.6 demonstrate the trajectories of four-lane freeway, where left-top, right-top, left-bottom and right-bottom sub-figures represents the trajectories on lanes 1, 2, 3 and 4, respectively. The y-axis of each sub-figure denotes the longitudinal location of vehicles, and the x-axis denotes the simulation

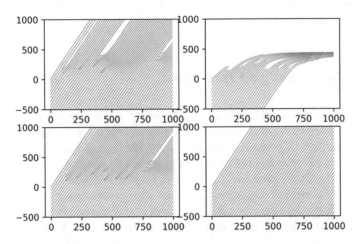

Fig. 17.2 Vehicle trajectories with reaction time and regular minimum acceptable gap

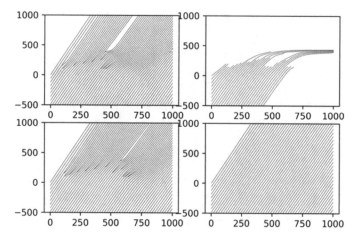

Fig. 17.3 Vehicle trajectories ignoring the reaction time

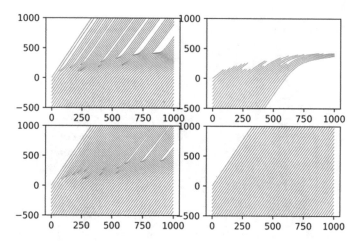

Fig. 17.4 Vehicle trajectories with reduced the minimum acceptable following gap

intervals whose unit is 0.1 s. As vehicles on lane 3 do not merge into lane 4, the trajectories on lane 4 can be regarded as undisturbed trajectories. When we apply regular lane-changing criteria that is illustrated in Eq (17.8), 8 out of 30 vehicles failed to execute the lane-changing motion before they stopped. Among the 24 successfully merged vehicles, 7 of them started lane-changing manoeuvres when their velocities were much lower than those of vehicles on the through lanes, which brought great impact to their surrounding vehicles, as shown in Fig. 17.2. When we improved the inter-vehicle communication by eliminating the negative effect of drivers' reaction time, only 3 out of 30 vehicles merged at relatively low speed, causing much less negative effect on the through lane; however, there were still 6 vehicles failed to merge as shown in Fig. 17.3. Alternatively, we reduced the $g_{f,min}$ by $\frac{V_{platoon}}{2}$ and depicted

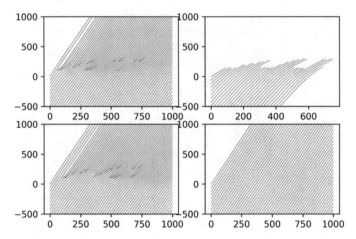

Fig. 17.5 Vehicle trajectories with reduced minimum acceptable following gap while ignoring the reaction time

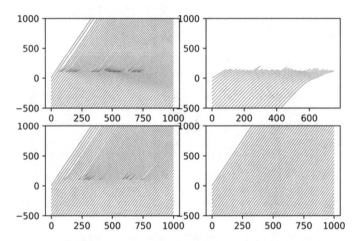

Fig. 17.6 Vehicle trajectories with ideal cooperation

the trajectories in Fig. 17.4. Even though the lane-changing motions seemed to be smooth, it brought much greater impact on the through lane, and that might be the result of over-aggressive lane-changing manoeuvres. In Fig. 17.5, we combined the two methods to smooth the lane-changing manoeuvres, results show a great improvement. This might be explained by the concept proposed by Hidas [5] that the large velocity difference can increase the lane changing. In our case, a smaller acceptable gap can encourage the driver to merge before velocity difference boost, so vehicles are able to merge without disturbing the through lanes. If we reduce the acceptable gap to the minimum requirement, almost all of the vehicles are able to commence merging immediately after drivers being inform of the occurrence of incidents as demonstrated in Fig. 17.6. It is also found that solely reduce the preceding or following minimum gap requirement cannot improve the performance of the traffic system, thus, the inter-vehicle communication with the preceding vehicles and following vehicles of target lanes are equally important.

17.4   Conclusion

In this research, we develop a four-lane traffic dynamic model and microscopically demonstrate the impact brought by bottlenecks on the freeway. One of the remarkable improvements is that our model incorporates a realistic FVD-based lane-changing trajectory by approximating the clothoids into cubic polynomials. Therefore, the time span of lane-changing manoeuvre should be equal to one simulation interval, instead, more realistic time spans are determined by the velocities and lane-changing trajectories of vehicles. Moreover, the inter-vehicle communication is indicated by the reduction of minimum acceptable gap, because inter-vehicle communication can bring connected and automated vehicles and drivers the confidence when executing lane-changing manoeuvres. It is found that the communication with the preceding and the following vehicle of target lane are equally important when we plan to improve the overall traffic performance. However, there are some more in-depth researches yet to be done, for instance, how cooperation among vehicles function need to be further researched. Additionally, the cooperative lane change when the following vehicle positively decelerate to encourage lane change are to be discussed, because the mechanism of our current model is based on free lane change and force lane change.

References

1. Bando, M., Hasebe, K., Nakayama, A., Shibata, A., Sugiyama, Y.: Dynamical model of traffic congestion and numerical simulation. Phys. Rev. E **51**, 1035 (1995)
2. Gackstatter, C., Heinemann, P., Thomas, S., Rosenhahn, B., Klinker, G.: Fusion of clothoid segments for a more accurate and updated prediction of the road geometry. In: International IEEE Conference on Intelligent Transportation Systems. IEEE (2010)
3. Helbing, D., Tilch, B.: Generalized force model of traffic dynamics. Phys. Rev. E **58**, 133 (1998)
4. Hidas, P.: Modelling lane changing and merging in microscopic traffic simulation. Transp. Res. Part C Emerg. Technol. **10**(5), 351–371 (2002)
5. Hidas, P.: Modelling vehicle interactions in microscopic simulation of merging and weaving. Transp. Res. Part C **13**(1), 37–62 (2005)
6. Jiang, R., Wu, Q., Zhu, Z.: Full velocity difference model for a car-following theory. Phys. Rev. E **64**, 017101 (2001)
7. Kasper, D., Weidl, G., Dang, T., Breuel, G., Tamke, A., Wedel, A., Rosenstiel, W.: Object-oriented Bayesian networks for detection of lane change maneuvers. IEEE Intell. Transp. Syst. Mag. **4**(3), 19–31 (2012)

8. Laval, J.A., Daganzo, C.F.: Lane-changing in traffic streams. Transp. Res. Part B Methodol. **40**(3), 251–264 (2006)
9. Lv, W., Song, W.G., Fang, Z.M.: Three-lane changing behaviour simulation using a modified optimal velocity model. Phys. A **390**(12), 2303–2314 (2011)
10. Meng, Q., Weng, J.: Cellular automata model for work zone traffic. Transp. Res. Rec. J. Transp. Res. Board **2188**, 131–139 (2010)
11. Qu, X., Wang, S., Zhang, J.: On the fundamental diagram for freeway traffic: A novel calibration approach for single-regime models. Transp. Res. Part B **73**, 91–102 (2015)
12. Sledge, N.H., Marshek, K.M.: Development and validation of an optimized emergency lane-change trajectory (No. 980231). SAE Technical Paper (1998)
13. Tang, T., Shi, W., Shang, H., Wang, Y.: A new car-following model with consideration of inter-vehicle communication. Nonlinear Dyn. **76**(4), 2017–2023 (2014)
14. Yu, Y., Jiang, R., Qu, X.: A modified full velocity difference model with acceleration and deceleration confinement: calibrations, validations, and scenario analysis. IEEE Intell. Transp. Syst. Mag. (2018)
15. Zheng, Z.: Recent developments and research needs in modeling lane changing. Transp. Res. Part B Methodol. **60**(1), 16–32 (2014)
16. Zou, Y., Qu, X.: On the impact of connected automated vehicles in freeway work zone: a cooperative cellular automata model based approach. J. Intell.Connect. Veh. **1**(1), 1–14 (2018)

Chapter 18
Friendliness Analysis for Bike Trips on Urban Roads Using Logistic Regression Model

Huichan Li, Zhiju Chen, Xiaohui Li and Yadan Yan

Abstract With rapid urbanization, problems such as traffic congestion and air pollution have become increasingly serious in urban areas. Therefore, governments and transport planners in Chinese cities are devoting efforts to implement urban development policies and encourage the use of non-motorized modes (walking and cycling). Mode share of biking trips may be retained and possibly increased if safe and convenient urban road facilities for users are created. Therefore, this paper utilizes the survey data from Zhengzhou city to examine the relationship between the attributes of road infrastructure and the user-perceived friendliness of cycling, by using the logistic regression model. Results indicate that urban road grades, types of barriers, lane width, etc., have important impacts on the friendliness of the biking mode.

18.1 Introduction

Contradiction between the rapid growth of the number of vehicles and limited road resources is increasingly prominent with the development of urbanization. More and more studies identified that traffic congestion problem was becoming more and more complicated and difficult to solve [1, 2]. Besides traditional traffic engineering measures [3, 4] and encouraging the use of public transport [5], many new technical methods such as connected vehicles and automated vehicles are also used in existing studies [6, 7].

H. Li · X. Li · Y. Yan (✉)
School of Civil Engineering, Zhengzhou University, Zhengzhou 450001, China
e-mail: yanyadan@zzu.edu.cn

H. Li
e-mail: 837195144@qq.com

X. Li
e-mail: 1195868149@qq.com

Z. Chen
School of Transportation & Logistics, Dalian University of Technology, Dalian 116024, China
e-mail: 512435575@qq.com

© Springer Nature Singapore Pte Ltd. 2019
X. Qu et al. (eds.), *Smart Transportation Systems 2019*, Smart Innovation,
Systems and Technologies 149, https://doi.org/10.1007/978-981-13-8683-1_18

Under this background, choosing bikes as trip modes will be more effective and convenient when the travel distance is not very long. Bike trips play positive roles in a sustainable and green transport system [8, 9]. Building a friendly biking environment is significant for stimulating meaningful changes in travel behaviour—increase biking and reduce car use. Hence, this paper analysed factors related to urban road infrastructure that affected the friendliness of bike trips. Based on survey data of several urban roads in Zhengzhou city, a bike friendliness prediction model of urban roads based on binary Logistic regression was then established.

18.2 Data Collection and Questionnaire

In this paper, infrastructure parameters of urban roads affecting the friendliness of bike trips were collected in two areas of Zhengzhou city, i.e. old district and high-tech district. The survey was conducted from 1 July to 1 August 2016. The data of 74 urban road sections in high-tech district and 98 urban road sections in old district were collected by field investigation, as shown in Fig. 18.1. Data including road grade, the lane width for bikes, consistency of bike lane width, types of barriers, pavement, the type and width of separate belt between the bike lane and motor vehicle lanes were all recorded.

Data collected from all urban road sections were sorted out and numbered. Then, an online questionnaire was formulated to enable respondents to determine the bike friendliness of each urban road section, according to their perception through pictures and videos. The questionnaire is shown in Fig. 18.2.

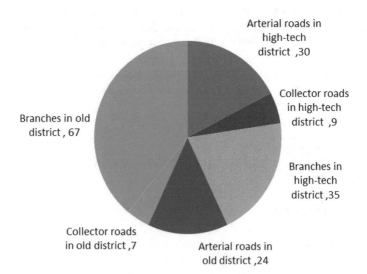

Fig. 18.1 The number of surveyed urban road sections in two districts

No. 23, Jinsuo road section (From Yinxing road to Yingchun street)	
	**Simple description:** · There are two motor vehicle lanes in two-way; · The separation type between the motor vehicle lane and bike lane is solid white marking; · The width of bike lane is 2 m; · The problem of parking on the bike lane exists in some places.
Please tick " √ " in the brackets if you feel that the investigation urban road section is friendly for bikes. Otherwise, cross "×" in the brackets. <div align="center">Friendly for bike trips ( )</div>	

Fig. 18.2 Sample of the designed questionnaire

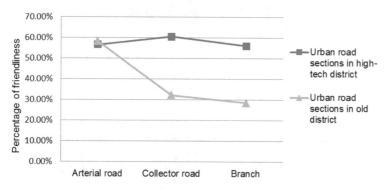

Fig. 18.3 The friendliness of the biking mode of Zhengzhou City

A total of 13,028 questionnaires were collected. Analysis results of bike friendliness for different grades of urban roads in high-tech district and old district are shown in Fig. 18.3. The percentage of arterial road sections in high-tech district which are friendly for bike trips is about 58%; the percentage of arterial road sections in old district which are friendly for bike trips is similar. As for collector roads and branches, the percentage values vary widely. Obviously, since the high-tech district is a new area, the urban road infrastructure is more friendly than that in old district.

18.3 Friendliness Analysis Based on Logistic Regression Model

18.3.1 Variables Selection

(1) Dependent Variables

The objective of modelling is to construct a predictive model for identifying the bike friendliness of the urban road section. Therefore, the friendliness for bike trips of the urban road section is the dependent variable, denoted by y. $y = 1$ suggests that the urban road section is friendly for bike trips and $y = 0$ suggests that the urban road section is unfriendly for bike trips.

(2) Independent Variables

According to the field investigation, 25 influential factors are selected as independent variables, as shown in Table 18.1. Among them, X_2, X_3, X_4, X_6, X_7, X_9, X_{13}, X_{16} and X_{17} are continuous variables, which are the actual values of field investigation, while other variables are categorical variables.

18.3.2 Logistic Regression Model

The biking friendliness prediction model for urban road sections is established by using the binomial logistic regression model. Factors of urban road sections affecting the friendliness of biking trips are analysed by logistic stepwise regression. Factors that have significant influence on the friendliness of biking trips are screened out by significant values, which indicate the significance of variables. Significant values less than 0.05 has statistical significance. The logistic regression model is established as follows:

$$
\begin{aligned}
\Pr_{bike} = Exp\big(&-0.231 + 0.211x_7 + 1.219x_{8(3)} - 0.421x_{11(2)} - 1.361x_{11(3)} - 2.008x_{14(1)} \\
&- 1.191x_{14(2)} + 0.638x_{14(3)} + 0.871x_{15(1)} + 3.829x_{16} + 0.154x_{17} - 4.277x_{18(1)} \\
&-0.338x_{18(3)} + 0.268x_{24(2)}\big)/\big(1 + Exp\big(-0.231 + 0.211x_7 + 1.219x_{8(3)} - 0.421x_{11(2)} \\
&- 1.361x_{11(3)} - 2.008x_{14(1)} - 1.191x_{14(2)} + 0.638x_{14(3)} + 0.871x_{15(1)} + 3.829x_{16} \\
&+0.154x_{17} - 4.277x_{18(1)} - 0.338x_{18(3)} + 0.268x_{24(2)}\big)\big)
\end{aligned}
\tag{18.1}
$$

The accuracy of the model prediction is checked, with 50% as the dividing value. As shown in Table 18.2, the accuracy of bike friendliness prediction is 66.6%, which indicates the good prediction.

Calibration results of the biking friendliness model are shown in Table 18.3. Significant influencing factors are the type of barriers of the sidewalk (X_{18}), sidewalk (X_{16}), bike lane pavement type (X_{14}), bike barrier type (X_{11}), the type of separate belt of motor vehicle and bike(X_8), bike pavement smoothness (X_{15}), road grade

Table 18.1 Independent variables for modelling

Independent variable	Description	Variable value	Independent variable	Description	Variable value
X_1	Direction of the road	East–west = 1; North–south = 2	X_{14}	Bike lane pavement type	Brick = 1, concrete = 2, asphalt = 3, coloured asphalt = 4
X_2	Total width of motor vehicle lanes		X_{15}	Bike pavement smoothness	Well paved = 1, no = 2
X_3	Number of motor vehicle lanes		X_{16}	Sidewalk	Have = 1, no = 2
X_4	Width of medial strip		X_{17}	Sidewalk width	
X_5	Type of medial strip	Green belt = 1, isolation fence = 2, no = 3	X_{18}	Type of barriers of the sidewalk	Bike parking = 1, motor vehicle parking = 2, road occupation of kiosk etc. = 3, no = 4
X_6	Width of the lay-by		X_{19}	Type of land use outside the red line	Store = 1, wall = 2,else = 3
X_7	The width of separate belt of motor vehicle and bike		X_{20}	Type of the traffic lights	Dedicated = 1, shared by motor and non-motor vehicle = 2, shared by non-motor vehicle and pedestrian = 3, no = 4
X_8	The type of separate belt of motor vehicle and bike	Green belt = 1, road marking = 2, isolation fence = 3, no = 4	X_{21}	Bus stop	Have = 1, no = 2

(continued)

Table 18.1 (continued)

Independent variable	Description	Variable value	Independent variable	Description	Variable value
X_9	Width of bike lane		X_{22}	Type of bus stop	Linear stop = 1, semi bay bus stop = 2, bus bay stop = 3
X_{10}	Consistency of the width of bike lane	Consensus = 1, discord = 2	X_{23}	Stop position of the bus	Separation zone of motor vehicles and non-motor vehicles lanes = 1, non-motor vehicle lane = 2
X_{11}	Bike barrier type	Motor vehicle parking = 1, road occupation of kiosk etc. = 2, bike parking = 3, no = 4	X_{24}	Road grade	Arterial road = 1, collector road = 2, branch = 3
X_{12}	Travel divided blocks of non-motor vehicle lane	Have = 1, no = 2	X_{25}	Road area	High-tech zone = 1, old city = 2
X_{13}	Interval of travel divided blocks of non-motor vehicle lane				

Table 18.2 Prediction accuracy of the model

Classification	Friendliness judgment		Prediction accuracy(%)
	Unfriendly	Friendly	
Unfriendly	2312	979	70.30
Friendly	1112	1861	62.60
Total	–	–	66.60

Table 18.3 Calibration results

Independent variable	Coefficient	S.E.	Wals	Sig.	OR
X_7	0.211	0.057	13.906	0	1.235
$X_{8(3)}$	1.219	0.221	30.343	0	3.384
$X_{11(2)}$	−0.421	0.078	29.058	0	0.656
$X_{11(3)}$	−1.361	0.246	30.686	0	0.256
$X_{14(1)}$	−2.008	0.424	22.376	0	0.134
$X_{14(2)}$	−1.191	0.155	59.214	0	0.304
$X_{14(3)}$	0.638	0.106	36.463	0	1.892
$X_{15(1)}$	0.871	0.303	8.291	0.004	2.389
X_{16}	3.829	1.017	14.167	0	46.006
X_{17}	0.154	0.022	51.063	0	1.167
$X_{18(1)}$	−4.277	1.014	17.794	0	0.014
$X_{18(3)}$	−0.338	0.08	17.622	0	0.713
$X_{24(2)}$	0.268	0.1	7.124	0.008	1.307
Constant term	−0.231	0.253	0.834	0	0.794

(X_{24}), the width of separate belt of motor vehicle and bike (X_7) and sidewalk width (X_{20}). From calibration results, it can be found that the friendliness of the urban road section for bike trips is positively affected by the separate belt of motor vehicle and bike, the bike lane width, the pavement of bike lane, the sidewalk outside bike lane and the sidewalk width of the bike lane. Road occupation of obstacles (for example, the kiosks) and parking on the bike lane have negative impacts. Compared with asphalt, brick and cement pavement for bike lane has poor comfort and low friendliness. Compared with ordinary asphalt, colour asphalt has higher friendliness. The friendliness of collector roads is 1.307 times that of branches, which indicates that transportation engineers pay more attention to the friendliness design of collector roads.

18.4 Conclusions

This paper aims to explore features of urban roads which are friendly for bike trips. Results show that the friendliness can be increased by properly improving the traffic environment of bike lanes and pavement performance. Setting appropriate isolation facilities between motor vehicle lanes, bike lanes and sidewalks will guide their traffic flow reasonably, therefore reducing the interference and make bike trips more friendly.

Besides, questionnaire respondent results suggest that the friendliness percentage of urban road sections in high-tech district is higher than that in old district. Traffic engineers and government-related departments have recognized the importance of urban road infrastructure in encouraging the use of bike modes.

References

1. Bulteau, J.: Revisiting the bottleneck congestion model by considering environmental costs and a modal policy. Int. J. Sustain. Transp. 10(3), 180–192 (2016)
2. Wang, S., Djahel, S., et al.: Next road rerouting: a multiagent system for mitigating unexpected urban traffic congestion. IEEE Trans. Intell. Transp. Syst. 17(10), 2888–2899 (2016)
3. Qu, X., Wang, S.: Long distance commuter lane: a new concept for freeway traffic management. Comput. Aided Civ. Infrastruct. Eng. 30(10), 815–823 (2015)
4. Ren, L., Qu, X., Guan, H., Easa, S., et al.: Evaluation of roundabout capacity models: an empirical case study. J. Transp. Eng. 142(12), 04016066 (2016)
5. Yan, Y., Liu, Z., Meng, Q., et al.: Robust optimization model of bus transit network design with stochastic travel time. J. Transp. Eng. 139(6), 625–634 (2013)
6. Zhou, M., Qu, X., Jin, S.: On the impact of cooperative autonomous vehicles in improving freeway merging: A modified intelligent driver model based approach. IEEE Trans. Intell. Transp. Syst. 18(6), 1422–1428 (2017)
7. Li, X., Ghiasi, A., Xu, Z., et al.: A piecewise trajectory optimization model for connected automated vehicles: exact optimization algorithm and queue propagation analysis. Transp. Res. Part B Methodol. 118, 429–456 (2018)
8. Verma, M., Rahul, T.M., Reddy, P.V., et al.: The factors influencing bicycling in the Bangalore city. Transp. Res. Part A Policy Pract. 89, 29–40 (2016)
9. Ogilvie, D., Egan, M., Hamilton, V., et al.: Promoting walking and cycling as an alternative to using cars: systematic review. Br. Med. J. 329(7469), 763–766B (2004)

Chapter 19
Change in Commuters' Trip Characteristics Under Driving Restriction Policies

Peng Zhao, Xiangyu Zhao, Yinmin Qian and Yadan Yan

Abstract In order to alleviate traffic congestion and environmental pollution, many cities in China have gradually adopted traffic demand management strategies. Driving restriction policies of motorized vehicles are widely implemented. Zhengzhou City implemented the even–odd number restriction policy from 4 December to 31 December 2017. Thereafter, from 1 January 2018, the tail number of license plate restriction policy was utilized. Under this background, an online questionnaire about commuters' trip characteristics before and after the driving restriction is designed. Using the survey data, commuters' acceptance of driving restriction policies and the influence of restriction policies on their travel mode choices are investigated. Results show that under the even–odd number restriction policy, most commuters' travel modes are transferred to private electric bicycles or bicycles, buses or subways; while under the tail number of license plate restriction policy, most commuters' travel modes are transferred to cars-hailing service, buses or subways. Besides, most commuters prefer the driving restriction policy with two tail number of license plate per day. Because new energy vehicles are not included in driving restriction policies, some commuters tend to buy a second car with new energy.

P. Zhao · Y. Yan (✉)
School of Civil Engineering, Zhengzhou University, Zhengzhou 450001, China
e-mail: yanyadan@zzu.edu.cn

P. Zhao
e-mail: 1209319710@qq.com

X. Zhao
Talent Apartment Development and Management Center, Zhengzhou University, Zhengzhou 450001, China
e-mail: zxylovely@zzu.edu.cn

Y. Qian
Department of Architectural Engineering, City University of Zhengzhou, Zhengzhou 452370, China
e-mail: 11090125@qq.com

© Springer Nature Singapore Pte Ltd. 2019
X. Qu et al. (eds.), *Smart Transportation Systems 2019*, Smart Innovation, Systems and Technologies 149, https://doi.org/10.1007/978-981-13-8683-1_19

19.1 Introduction

With the rapid development of urbanization in China, the population in cities is more and more dense and the number of motorized vehicles is increasing quickly. In order to alleviate traffic congestion and environmental pollution problems, many measures and new technical methods are proposed in existing studies [1, 2]. Li et al. [3] proposed a piecewise trajectory optimization model for connected automated vehicles. Wang et al. [4] proposed a trial-and-error fare design scheme which will reduce traffic congestion at commuter train stations. Qu et al. [5] proposed a novel concept: long-distance-commuter lane. The lane was a dedicated lane that only commuter travelling between the two major cities could use. Moreover, the use of public transport is encouraged [6].

In addition, many cities began to adopt traffic demand management strategies. It is expected to reduce the amount of traffic volume on the urban road and the emissions from motorized vehicle using the policy and economic. Among these measures, motorized vehicle driving restriction is widely utilized. There have been some existing studies on driving restriction issues, which mainly focus on whether traffic congestion can be alleviated and air quality can be improved. Using family travel survey data, Guerra et al. [7] found that a driving restriction policy had little effects on reducing overall vehicle travel, because commuters changed their travel plans. Taking Tianjin city as an example, Jia et al. [8] and Liu et al. [9] studied the acceptance degree and behaviour response of commuters to the policy by questionnaire investigation and mathematical models. Results showed that the driving restriction policy alone cannot effectively encouraged vehicle owners to use public transport. Sun et al. [10] analyzed the marginal effect of driving restriction policies in Beijing, It was indicated that effects of driving restriction on alleviating traffic congestion were significant, but the effect on reducing respirable particulates was not obvious; Wang et al. [11] studied household surveys and travel diaries, suggesting that driving restriction in Beijing had no significant impacts on individual driving decisions, because violations of the driving restriction rule were persistent and widespread.

Zhengzhou city has continuously implemented two different restriction policies in a short period of time, including the even–odd number restriction policy and the tail number of license plate restriction policy. Under this background, this paper carried out an investigation on changes of commuters' trip characteristics and acceptance of driving restriction policies.

19.2 Driving Restriction Policies in Zhengzhou City

19.2.1 Even–Odd Number Restriction Policy

The driving restriction area includes all roads within the third ring expressway of Zhengzhou City. The driving restriction period was from 0:00 o'clock on 4 December 2017 to 24:00 on 31 December 2017.

According to the last number of motorized vehicle license plate, cars with odd numbers of license plates (1, 3, 5, 7, 9) were permitted to drive on urban roads within the driving restriction area on single dates. Cars with even numbers of license plates (2, 4, 6, 8, 0) were permitted to drive on double dates. New energy cars were not restricted by the policy.

19.2.2 Tail Number of License Plate Restriction Policy

The driving restriction area is the same as that of the even–odd number driving restriction. The driving restriction period is from January 1 to November 20 2018, during 7:00–21:00 on weekdays. According to the last number of license plates, cars with two specific tail numbers were restricted to drive on urban roads per day. Cars whose last Arabic number was 1 or 6 were restricted on Monday; cars whose last Arabic number was 2 or 7 were restricted on Tuesday; cars whose last Arabic number was 3 or 8 were restricted on Wednesday; cars whose last Arabic number was 4 or 9 were restricted on Thursday; cars whose last Arabic number was 5 or 0 were restricted on Friday. Also, new energy cars were not restricted by the policy.

19.3 Questionnaire Design and Survey Data Analysis

19.3.1 Questionnaire Design

In order to analyze the acceptance of motorized vehicle driving restriction policies and the influence of motorized vehicle restriction policies on choices of commuters' travel modes, a questionnaire was designed. The questionnaire contained 22 questions which could be classified into 5 aspects:

- attributes of commuters, such as gender, age, the number of owned cars, etc.;
- origin and destination of commuting trips;
- travel modes before and after driving restriction policies;
- attitude towards driving restriction policies, such as whether you think they could contribute to the improvement of traffic congestion or air quality, etc.;
- attitude towards new energy vehicles which are not restricted.

19.3.2 Changes of Commuting Trip Modes

A total of 352 questionnaires were collected online. Results of cross analysis between trip modes and commuters' attributes before driving restriction policies were shown in Table 19.1. Before the implementation of driving restriction policies, 67.6% of commuters use private cars as daily trip modes. The second type of trip mode that is often used by commuters is private electric bicycles or bicycles, accounting for 11.6%. For men, 72.4% use private cars and 11.0% use private electric bicycles or bicycles. While for women, only 55.1% use private cars.

Cross-analyses of the even–odd number and tail number of license plate driving restriction policies are shown in Tables 19.2 and 19.3. After the even–odd number driving restriction policy was implemented, the proportion of private cars decreased from 67.6 to 27.5%. The use of public transport increased greatly, from 8.5 to 33.9%. The proportion of private electric bikes or bicycles showed almost no change. After the tail number of license plate driving restriction policy was implemented, the proportion of car travel decreased to 30.6%. The proportion of bus and subway travel was 27.8%, followed by the proportion of private electric vehicles or bicycles, which increased to 16.7%.

19.3.3 Acceptance of Driving Restriction Policies

According to Tables 19.4 and 19.5, 53.7% of commuters think that driving restriction policies help to improve air quality. 70.4% of commuters agree that driving restriction policies are helpful for encouraging the use of public transport.

As shown in Tables 19.6 and 19.7, if restrictions are implemented for a long period, about 59.4% of commuters consider buying a second car. Besides, since new energy vehicles are not restricted by the policy, also about 59.1% of commuters consider buying new energy vehicles.

It can be found in Table 19.8 that 34.4% of commuters choose weekdays as the implementation period of driving restriction policies. 22.7% of commuters choose long-time as the implementation period of driving restriction policies. There are also 33.5% of commuters are against the implementation of driving restriction policies.

Attitudes on different types of driving restriction policies are shown in Table 19.9. Compared with the even–odd number restriction policy and one tail number per day restriction policy, commuters prefer the two tail numbers per day driving restriction policy.

Table 19.1 Proportion of commuting modes before driving restriction policies

The number of questionnaire	Trip modes								
	Private car	Cars-hailing service	Private electric bicycle or bicycle	Bicycle sharing	Bus or subway	Car sharing	Taxi	Carpooling with colleagues	Walk
Male (people)	184	2	28	7	17	0	0	4	12
Female (people)	54	4	13	1	13	0	0	2	11
Proportion (%)	67.6	1.7	11.6	2.3	8.5	0.0	0.0	1.7	6.5

Table 19.2 Proportion of commuting modes after the even–odd number driving restriction policy

The number of questionnaire	Trip modes								
	Private car	Cars-hailing service	Private electric bicycle or bicycle	Bicycle sharing	Bus or subway	Car sharing	Taxi	Carpooling with colleagues	Walk
Male (people)	68	14	25	10	77	1	9	9	16
Female (people)	19	6	15	2	30	0	1	6	8
Proportion (%)	27.5	6.3	12.7	3.8	33.9	0.3	3.2	4.7	7.6

Table 19.3 Proportion of commuting modes after tail number of license plates driving restriction policy

The number of questionnaire	Trip modes								
	Private car	Cars-hailing service	Private electric bicycle or bicycle	Bicycle sharing	Bus or subway	Car sharing	Taxi	Carpooling with colleagues	Walk
Male (people)	8	0	5	1	7	0	1	0	3
Female (people)	3	0	1	0	3	0	0	1	3
Proportion (%)	30.6	0.0	16.7	2.8	27.8	0	2.8	2.8	16.7

Table 19.4 Attitude on whether driving restrictions help improve air quality

Gender	Attitudes		Total
	Yes	No	
Male (people)	128	126	254
Female (people)	61	37	98
Total	189	163	352
Proportion (%)	53.7	46.3	100

Table 19.5 Attitude on whether driving restrictions help encourage the use of public transport

Gender	Attitudes		Total
	Yes	No	
Male (people)	169	85	254
Female (people)	79	19	98
Total	248	104	352
Proportion (%)	70.4	29.6	100

Table 19.6 Commuters' attitude on buying a second car

Gender	Attitudes		Total
	Yes	No	
Male (people)	170	84	254
Female (people)	39	59	98
Total	209	143	352
Proportion (%)	59.4	40.6	100

Table 19.7 Commuters' attitude on buying new energy vehicles

Gender	Attitudes		Total
	Yes	No	
Male (people)	152	102	254
Female (people)	56	42	98
Total	208	144	352
Proportion (%)	59.1	41.9	100

Table 19.8 Time period of implementing driving restriction policies

Gender	Implementation time					Total
	Holidays	Weekends	Weekdays	Long term	Non-restrictions	
Male (people)	12	10	92	51	89	254
Female (people)	5	6	29	29	29	98
Total	17	16	121	80	118	352
Proportion (%)	4.8	4.5	34.4	22.7	33.5	100

Table 19.9 Attitude on different types of driving restriction policies

Gender	Types of driving restriction policies			Total
	Even–odd	One tail number per day	Two tail numbers per day	
Male (people)	56	59	139	254
Female (people)	30	29	39	98
Total	86	88	178	352
Proportion (%)	24	25	51	100

19.4 Conclusions

By designing questionnaires and conduct statistics, this paper analyses changes in commuters' trip characteristics under different driving restriction policies. After restriction policies were implemented, original car trips are transferred to public transport and electric bikes or bicycles. More than half commuters think that driving restriction policies help to improve air quality. Compared with women, men are more positive about buying a second car and new energy vehicles.

References

1. Zhou, M., Qu, X., Jin, S.: On the impact of cooperative autonomous vehicles in improving freeway merging: a modified intelligent driver model based approach. IEEE Trans. Intell. Transp. Syst. **18**(6), 1422–1428 (2017)
2. Zhou, M., Qu, X., Li, X.: A recurrent neural network based microscopic car following model to predict traffic oscillation. Transp. Res. Part C Emerg. Technol. **84**, 245–264 (2017)
3. Li, X., Ghiasi, A., Xu, Z., et al.: A piecewise trajectory optimization model for connected automated vehicles: exact optimization algorithm and queue propagation analysis. Transp. Res. Part B Methodol. **118**, 429–456 (2018)
4. Wang, S., Zhang, W., Qu, X.: Trial-and-error train fare design scheme for addressing boarding/alighting congestion at CBD stations. Transp. Res. Part B Methodol. **118**, 318–335 (2018)
5. Qu, X., Wang, S.: Long distance commuter lane: a new concept for freeway traffic management. Comput. Aided Civ. Infrastruct. Eng. **30**(10), 815–823 (2015)
6. Yan, Y., Liu, Z., Meng, Q., et al.: Robust optimization model of bus transit network design with stochastic travel time[J]. J. Transp. Eng. **139**(6), 625–634 (2013)
7. Guerra, E., Millard-Ball, A.: Getting around a license-plate ban: behavioral responses to Mexico City's driving restriction. Transp. Res. Part D Transp. Environ. **55**, 113–126 (2017)
8. Jia, N., Zhang, Y.D., He, Z.B., et al.: Commuters' acceptance of and behavior reactions to license plate restriction policy: a case study of Tianjin, China. Transp. Res. Part D Transp. Environ. **52**, 428–440 (2017)
9. Liu, Y.X., Hong, Z.S., Liu, Y.: Do driving restriction policies effectively motivate commuters to use public transportation? Energy Policy **90**, 253–261 (2016)

10. Sun, C., Zheng, S.Q., Wang, R.: Restricting driving for better traffic and clearer skies: did it work in Beijing? Transp. Policy **32**, 34–41 (2014)
11. Wang, L.L., Xu, J.T., Qin, P.: Will a driving restriction policy reduce car trips?—the case study of Beijing, China. Transp. Res. Part A Policy Pract. **67**, 279–290 (2014)

Chapter 20
Simulated CAVs Driving and Characteristics of the Mixed Traffic Using Reinforcement Learning Method

Jingqiu Guo, Yangzexi Liu and Shouen Fang

Abstract Using cooperative gaming method can be a promising approach to mimic various driving tasks in the field of automated driving. This paper presents a Deep Reinforcement Learning approach for modelling Connected and Automated Vehicles (CAVs) in heterogeneous traffic. First, the Gipps models were integrated into regular vehicle agent. Second, an enhanced Q-learning was employed as the modelling platform for CAVs, to strengthen the capability of the simulation system in realistically reproducing CAV lane-changing and car-following behaviour. Third, extensive simulation studies based on a two-lane highway stretch show that the inclusion of CAVs considerably improves traffic flow, mean speed, and traffic capacity. We also simulated managed lane policies to determine how CAVs should be distributed across lanes in various conditions. Such understanding is essential for research concerning CAV, as well as, the CAV implication for future traffic management.

20.1 Introduction

The advanced capabilities of Connected and Autonomous Vehicles (CAVs) provide enormous opportunities to improve traffic safety, efficiency and sustainability [1]. However, for a long period, the market penetration of CAVs will gradually increase, and so CAVs will share limited road networks with Regular Vehicles (RVs). Traffic flow effects of CAVs may be examined in multiple aspects such as travel time, traffic capacity, flow stability, shockwave mitigation, etc For example, in a mixed context with both CAVs and RVs, the smaller a mean time gap could be maintained by CAVs, the bigger an achievable highway capacity would result [2]. On the other

J. Guo · Y. Liu (✉) · S. Fang
Key Laboratory of Road and Traffic Engineering of the Ministry of Education, Tongji University, 200092 Shanghai, China
e-mail: 16522@tongji.edu.cn

J. Guo
e-mail: guojingqiu@hotmail.com

S. Fang
e-mail: 18508230695@163.com

© Springer Nature Singapore Pte Ltd. 2019
X. Qu et al. (eds.), *Smart Transportation Systems 2019*, Smart Innovation, Systems and Technologies 149, https://doi.org/10.1007/978-981-13-8683-1_20

hand, CAVs have the ability to communicate with surrounding vehicles of the same type, which enables CAVs to gain more information during car-following and lane-changing operations, implement more flexible and intelligent decisions [3]. But still one cannot conclude clear conclusions about their impact on traffic performance [4].

A considerable body of literature is available on the topic of the intelligent vehicle driving models. A common approach to the model of automated vehicles is to integrate intelligent lateral and longitudinal driving decisions. For example, Intelligent Driver Model (IDM), formulated as an ordinary differential equation treating space and time as continuous variables [5], has been found to be a valid basis for the modelling of automated vehicles, and extended IDMs have been implemented in real automated vehicle testing. Lane-change Model with Relaxation and Synchronization (LMRS), incorporates relaxation where small headways are accepted during lane changes, and synchronization as a form of lane-change preparation [6].

Machine learning algorithms have made great breakthroughs in natural language processing, image recognition, behaviour, e.g. [7, 8]. Reinforcement Learning (RL) is one promising branch in machine learning family and is able to deal with unforeseen problems and seeking optimal solutions for long-term objectives via learning from the process of trial and error [9]. In recent years, it has been applied in automated vehicle's decision-making [10]. Q-learning is a popular model-free RL method, which approaches the optimal strategy in Markov decision-making process by iterations, and can well reflect the uncertainty and intelligence of CAV driving behaviour. Generally, three key elements of deep RL are: (1) proper RL algorithm, (2) a variety of training environment to improve the generalization, (3) reward function design. However, the existing RL studies on CAVs have some limitations in these aspects.

Automated vehicle's behaviour decision-making deserves in-depth investigations by considering both demands and constraints. In this paper, we will focus on the aspect of traffic flow characteristic analysis and lane policies. First, the well-developed Gipps models are used for RVs, and RL is proposed to train the CAV agents such that CAVs can be more agile on the manoeuvres of lane-changing and car-following. Particularly, we designed an enhanced deep Q-learning algorithm that has a closed-form greedy policy, which contributes to the computation efficiency. Second, extensive simulations are conducted with respect to a two-lane highway stretch, and the results illustrate that the RL based method is capable of learning efficient and sophistic driving scenario, in consideration of a variety of market penetration rate of CAVs. Third, we discussed CAV dedicated lane policies.

20.2 Methodology

20.2.1 Policies for RVs

We first integrate Gipps model into the RV agent. Regardless of the vehicle type of the front vehicle, let $d_{safe,n}$ denote the minimal safety-spacing between vehicle n and its front vehicle $n - 1$ on the same lane:

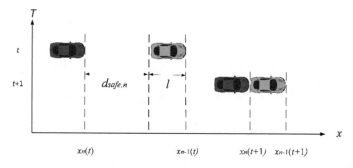

Fig. 20.1 The schematic diagram of safe car-following gap

$$d_{safe,n} = x_{n-1}(t) - x_n(t) - l = \mu \cdot v_n(t) + \frac{1}{2b}\left(v_n(t)^2 - v_{n-1}(t)^2\right) \qquad (20.1)$$

$$d_{safe,n}(t) = \lceil \frac{d_{safe,n}}{l_{cell}} \rceil \cdot l_{cell} \qquad (20.2)$$

where $x_n(t)$ is the position of the vehicle of n at time t, l is the length of the vehicle $n-1$, μ is the driver's reaction time, b is the maximal deceleration value, and l_{cell} is the cell size. This criterion of $d_{safe,n}$ at time t ensures that vehicle n cannot collide with its front vehicle $n-1$, as shown in Fig. 20.1.

$$v_{safe,n}(t+1) = \min(\lfloor \left\{ v_n(t) + \frac{2.5a\mu}{l_{cell}}\left[1 - \frac{v_n(t)}{v_{max}}\right] \cdot \sqrt{0.025 + \frac{v_n(t)}{v_{max}}} \right\} / l_{cell} \rfloor,$$

$$\lfloor \left(\mu b + \sqrt{(\mu b)^2 - b\left\{ 2[x_{n-1} - x_n - l] - \mu v_n(t) - \frac{2v_{n-1}(t)^2}{b_{n-1}(t) + b_{n-1}(t-1)} \right\}} \right) / l_{cell} \rfloor) \qquad (20.3)$$

where a denotes the maximal acceleration value, v_{max} is the maximal vehicle speed, $b_{n-1}(t)$ is the deceleration value of the front vehicle at t. $v_{safe,n}(t+1)$ is the maximal speed of vehicle n at t to promise the safety-spacing between vehicle n and its front vehicle $n-1$ on the same lane at time $t+1$. The updating rules followed classic NaSch model [11] to emulate step-wise vehicle motions, and the detailed formulations can be found in [12].

20.2.2   Policies for CAVs

20.2.2.1   Q-learning-Based Training Model

We designed an enhanced deep Q-learning algorithm that has a closed-form greedy policy, which contributes to the computation efficiency and stability. Q-learning consists of three aspects: (1) Environment (E); (2) Action (A); (3) Reward (r) [13]. The process of choosing a specific action A under the state S is the policy π of an agent, that is, $\pi : S \rightarrow A$. Thus, the agent will firstly decide an action a at time t (the state s_t), then the environment will send the reward to this agent as the feedback, which is used to evaluate the value of a and to decide the next action at the next state s_{t+1}. The accumulated reward V is the discounted value of the future total reward, where the discounted factor is γ $(0 \leq \gamma \leq 1)$. The agent attempts to maximize V, and the feedback reward system will lead it to take actions at the continual time dimension with much artificial intelligence in retrospective. $Q(s, a)$ is the function approximation of a state–action pair, and it can be calculated as

$$V^{\pi}(S_t) = \sum_{j=0}^{\infty} \gamma^j r_{t+j} \tag{20.4}$$

$$\pi^* = \underset{\pi}{\text{argmax}} \, V^{\pi}(s) \tag{20.5}$$

$$Q^{\pi}(s, a) = r(s, a) + \gamma \max_{a'} Q\big(\delta(s, a), a'\big) = E^{\pi} \left[\sum_{t=0}^{\infty} \gamma^t r_t | s_0 = s, a_0 = a \right] \tag{20.6}$$

where j is the relative approaching time point compared to the time point t, and $\delta(s, a)$ is the state transforming function. Naturally, the update of $Q(s, a)$ is set to satisfy Bellman Equation:

$$Q^{\pi}(s_t, a_t) = \sum_{s_{t+1}} \big[p(s_t, a_t, s_{t+1}) \cdot r(s_t, a_t, s_{t+1}) \big]$$

$$+ \gamma \sum_{s_{t+1}, a_{t+1}} \big[p(s_t, a_t, s_{t+1}) \cdot Q^{\pi}(s_{t+1}, a_{t+1}) \big] \tag{20.7}$$

where $p(s_t, a_t, s_{t+1})$ is the probability of the agent's state transforming $((s_t, a_t) \rightarrow s_{t+1})$, and $r(s_t, a_t, s_{t+1})$ is the reward value during that process. The optimal function approximation $Q^{\pi^*}(s, a)$ and the optimal policy $\pi^*(s)$ at state s are as follows:

$$Q^{\pi^*}(s, a) = \max_{\pi} Q^{\pi}(s, a) \tag{20.8}$$

$$\pi^*(s) = \underset{\varpi}{\text{argmax}} \big[r(s, a) + \gamma V^*(\delta(s, a)) \big] = \underset{a}{\text{argmax}} \, Q(s, a) \tag{20.9}$$

20.2.2.2 The Definition of the Vehicle's State

With the ability of sensing the surrounding environment, CAV n should at least consider the driving states of itself, the proceeding vehicle $n - 1$ at the same lane, the proceeding and succeeding vehicles $n - 2, n + 2$, respectively while conducting car-following and lane-changing processes. Thus, in this study we consider the above four vehicle states to determine a CAV's driving decision. In order to simulate more realistic situations, vehicle n needs to consider states over a series of timesteps and then chooses an optimal driving policy.

Based on the definition of a CAV's state, we designed the Q-table where the vertical axis and horizontal denote CAV's state and action. For simplicity, we just consider that the valid longitudinal sensing scale of a CAV equals to $2v_{max}$ separately compared to itself. Thus, the state S of vehicle n at time t can be represented by a 10-dimensional vector

$$S_n(t) = \left[v_{n+2}; p_{n+2}; d_{n+1,other}; v_n; d_n; d_{n,other}; v_{n-1}; p_{n-1}; v_{n-2}; p_{n-2} \right] \quad (20.10)$$

where p_i is the type of the vehicle at the relative location i ($i \in \{n-1, n-2, n+2\}$, $p_i \in \{CAV, RV, None\}$). Furthermore, when $p_i = None$, then $v_i = 0$.

20.2.2.3 The Choose of Action

Generally, a vehicle's action space consists of total six actions, which are (1) decrease the speed at current lane "F−"; (2) keep the speed at current lane "F="; (3) accelerate the speed at current lane "F+";(4) decrease the speed while lane changing "C−"; (5) keep the speed while lane changing "C="; (6) accelerate the speed while lane changing "C+". To optimize the training process as well as to avoid collision, some priori knowledge has been added. For example, when $d_n = 0$, the vehicle cannot conduct "F+" indeed.

$A_{feasible,n}(S)$ describes the feasible action space of CAV n, where $A_{feasible,n}(S) \in A_{all}$. We employ $\epsilon -$ greedy policy to determine which action the vehicle will take, where the vehicle is supposed to choose the specific action with the maximal action value in a ϵ probability, while the vehicle will randomly choose an action with a $(1 - \epsilon)$ probability,

$$\pi(S) = \begin{cases} \underset{a}{\operatorname{argmax}} \, Q\left(S, A_{feasible,n}(S)\right), & rand() \leq \epsilon \\ F\left(A_{feasible,n}(S)\right), & else \end{cases} \quad (20.11)$$

where $rand()$ denotes a random number in $[0, 1]$, $F(\cdot)$ is the random choosing function. The principle of reward is connected to the driving objective that every vehicle is assumed to gain the maximal speed through the whole simulation process, as $r = v_n(S') - v_n(S)$, where $v_n(S)$ is the speed of vehicle n at state S, and $S' : S \times \pi(S)$.

20.2.2.4 Hybrid Training

In consideration of effectiveness and stability, we replace the episode mechanism of classic Q-learning, and merge the concepts of discrete timestep for RVs and iteration of Q-learning for CAVs. Meanwhile, we apply a shared Q-table pool for all CAVs in the simulation, a way to significantly accelerate training.

20.3 Simulation and Results

We developed the simulation environment in Python. A 3 km two-lane highway stretch as illustrated in Fig. 20.2 is considered for the stimulation. The maximal speed for all vehicles in the study is 90 km/h, and the maximal acceleration value a and the maximal deceleration b are 5 m/s^2 and -10 m/s^2. The lane-changing capacity level $\delta = \{-2, -1, 0, 1, 2\}$. The length of each cell is equal to 1 m, so 5-cell size is approximately the average length of a vehicle. In this work, we focus on the case of the lane-changing probability $p_c = 1$ and so explore the deliciated impacts of CAVs on lane-wise lane changing. The number and locations of cells that are occupied with vehicles are randomly initialized at the start of each simulation run. As a routine practice, periodic boundary conditioning is considered. Let N, and β denote the total number of vehicles, the penetration rate of CAVs within the whole highway stretch. Denote by T, Q, $\bar{v}$ and ρ the simulation time horizon, the average flow, the average speed and the density of the whole highway. Denote by ρ_i the density and the length of ith lane of the stretch, and by N_i the number of vehicles included in the ith lane of the simulated stretch ($i = \{1, 2\}$). With these definitions, we have

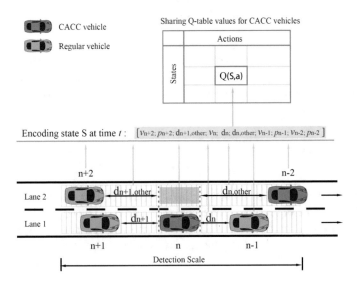

Fig. 20.2 The schematic diagram of CAV's state

$$\bar{v} = \frac{1}{T} \sum_{t=t_0}^{T+t_0-1} \frac{1}{N} \sum_{n=1}^{N} v_n(t) \tag{20.12}$$

$$\rho = \frac{N}{2L} \tag{20.13}$$

$$\rho_i = \frac{N_i}{L} \tag{20.14}$$

$$Q = \rho \cdot \bar{v} \tag{20.15}$$

The simulation platform includes both training runs and simulation runs. In training process, with regard to various CAV penetration rates and densities, it lasts 10^6 time steps to stabilize CAVs decision-making pattern. For each simulation run, the outcome from the first 15,000 simulation time steps was eliminated in order to remove transient effects, and the simulation outcome of the next 5000 time steps was considered for this work. The simulation was run 20 times for each scenario.

20.3.1 Analysis of Traffic Flow Characteristics

The obtained mixed traffic diagrams are presented in Fig. 20.3a, b. First, Fig. 20.3a displays the speed–density relation extracted from the simulations. For a certain CAV penetration rate β, the density feature is inversely correlated with the vehicle speed, when 30 veh/km $\leq \rho \leq$ 40 veh/km, the effect is most significant (Fig. 20.4).

On the other hand, the influence of β on speed shows non-linearity, that is, the evolution of CAVs under Q-learning is different from that of rule-based RVs. When ρ is in the range of 0–20 veh/km (sparse traffic flow), the influence of β on speed is negligible. For a ρ among 20–60 veh/km and β in the range of 0–0.65, the influence of penetration rate on the speed is weak, while the traffic flow still has a larger average

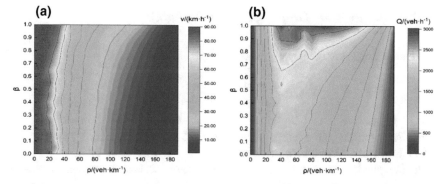

Fig. 20.3 Features of the mixed traffic caused by β and ρ. **a** Effects on the traffic velocity; **b** Effects on the traffic flow

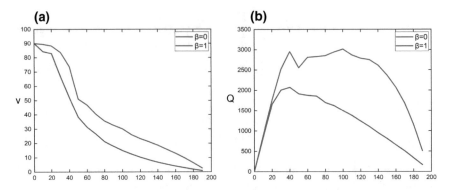

Fig. 20.4 Fundamental diagram of homogeneous traffic. **a** speed–density diagram; **b** flow density diagram

speed; with a greater β towards 1.00, the influence of β on speed is strengthened, which shows that under the same density, the higher β, the greater the speed of the traffic flow; in the range of 60–160 veh/km (the traffic flow is slightly congested to heavier congested state), β plays a significant role to keep speed at a high level with the increasing congestion. When ρ is more than 160 veh/km, traffic is in serious congestion, and the influence degree of β on vehicle flow speed is further reduced, but it still satisfies the positive correlation.

The road capacity is $Q_{max} = 2073$ veh/h at zero percentage of CAVs ($\beta = 0$) and becomes $Q_{max} = 3013$ veh/h when all vehicles are CAVs ($\beta = 1$), about 45.34% higher. Clearly, the involvement of CAVs increases the overall road capacity and average speed considerably. The results reflect that CAVs allow smaller longitudinal headways, more specifically, the fully optimized CAV agent via deep reinforcement learning evaluates each optional action in advance and can make optimal driving behaviour decisions in order to achieve greater speed in dynamic traffic environment.

20.3.2 Analysis of Lane-Changing Probability

The lane-changing frequency f_{LC} is defined as

$$f_{LC} = \sum_p f_{p,LC} = \sum_p \frac{N_{p,LC}}{T \cdot N_p} \tag{20.16}$$

where $N_{p,LC}$ refers to the total number of lane-changing instances, N_p denotes the total number of CAVs or RVs in the entire simulation, p is {CAV, RV}.

In this section we focus on the case of lane-wise lane changing, as shown in Fig. 20.5. On the one hand, f_{LC} $f_{CAV,LC}$ $f_{RV,LC}$ evolve along a fundamental-diagram-like trend over various β. Under different densities, the more the CAVs

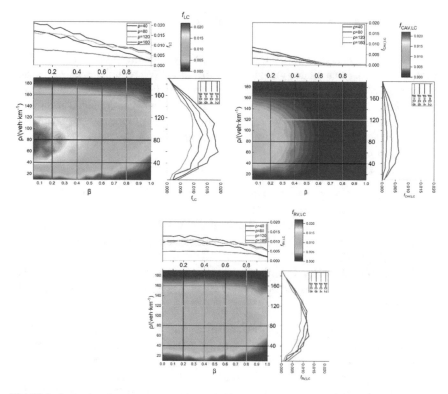

Fig. 20.5 Lane-changing frequency analysis of different β and ρ (top left: f_{LC}; top right: $f_{CAV,LC}$; bottom: $f_{RV,LC}$)

involved, the less the values of f_{LC}, $f_{RV,LC}$, $f_{CAV,LC}$, and the maximal lane-changing frequencies $f^*_{RV,LC} > f^*_{CAV,LC}$. This indicates that high-speed driving reduces the safe gap for RVs' lane-changing, so $f_{RV,LC}$ tends to decrease. Meanwhile, because CAVs can manage the car-following process more efficiently even with much smaller intervehicle spacing than RVs. Last, CAVs are able to decide the optimal driving behaviour in adjacent time steps and hence have less incentive to change lanes. On the other hand, according to Fig. 20.5, it is interesting to see the critical/turning points in terms of traffic densities (around 60 vehs/km). This indicates that the densities ρ of both lanes are low, vehicles have more freedom to do lane changing, and f_{LC} $f_{CAV,LC}$ $f_{RV,LC}$ increase with the number of vehicles on roads until reaching a turning point and then starts to decrease. After the peak, lane changing becomes less and less practical. The simulation results also suggest that the critical/turning points are more sensitive to the $f_{RV,LC}$ value.

20.4   Managed Lane Considerations

In the CAV era, managed lane design, such as lane orientation and separation, should be made to ensure that all design elements work to achieve regional consistency and efficiency. Different lane policies to better accommodate CAVs has received very little attention in the previous research [14, 15]. This work is intended to provide insights into the impact of CAV dedicated lanes on overall throughput considering CAV penetration rate, CAVs performance and traffic density, to reflect the gradually increasing introduction of CAVs.

Here we study the traffic flow under four different lane policies:

1. (Mixed, RV) policy, in which lane 1 allows both CAV platoons and RVs (mixed traffic), and lane 2 is for RVs only;
2. (CAV, RV) policy, in which lane 1 only allows CAV platoons and lane 2 for RVs;
3. (CAV, Mixed) policy, in which lane 1 is dedicated to CAV platoons and lane 2 has mixed traffic.
4. (Mixed, Mixed) policy, in which lane 1 and 2 both allow mixed traffic, in other words, there is no managed lane policy which has been discussed in the previous sections.

The first three policies have been simulated using reinforcement learning based on the two-lane highway stretch. We present some main outcomes in Fig. 20.6. The results cover a full density and penetration rate spectrum in order to form a comprehensive understanding of impacts of CAV managed lanes. Conclusively, we found that strict separation of CAVs and RVs ((CAV, RV) policy) can lead to underutilization of traffic capacity and the mixed-use ((Mixed, RV) or (CAV, Mixed)) policies can realize higher capacities. Changes of feasible lane policies also depend on β.

20.5   Conclusions

By exploring an evolutionary mechanism using an improved reinforcement learning method, an efficient and promising simulation method for the mixed traffic flow in two-lane environment is constructed. The following conclusions are: (1) Compared with rule-based models, RL is able to learn from interaction with a complex environment; (2) Traffic capacity and speed increase positively with the penetration rate of CAVs; (3) The lane-changing frequency between neighbouring lanes evolves with traffic density along a fundamental-diagram-like curve; (4) The dynamic relationship between CAV lane policies, CAV performance, penetration rate and density was revealed, which is helpful in deciding the optimal number of lanes to be allocated to CAVs in an evolving period. The proposed approach can be easily extended to various multi-lane models based on specific scenarios, indicating great potentiality for CAV studies in the near future.

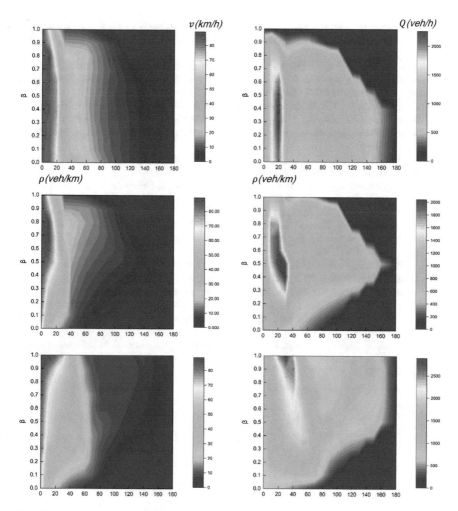

Fig. 20.6 Features of the traffic in various lane policies caused by β and ρ. (left) Effects on the traffic velocity; (right) Effects on the traffic flow. First line: M-R policy; Second line: A-R policy; Third line: M-A policy

References

1. Shladover, S.E.:. Connected and automated vehicle systems: introduction and overview. J. Intell. Transp. Syst. (2017)
2. Talebpour, A., Mahmassani, H.S., Hamdar, S.H.: Modeling lane-changing behavior in a connected environment: a game theory approach. Transp. Res. Part C (2015)
3. Khan, U., Basaras, P., Schmidt-Thieme, L., et al.: Analyzing cooperative lane change models for connected vehicles. In: 2014 International Conference on Connected Vehicles and Expo, pp. 565–570 (2014)

4. Anargiros, I.D., Nikolos, I.K., Papageorgiou, M.: A Macroscopic Multi-lane Traffic Flow Model for ACC/CACC Traffic Dynamic. Transportation Research Board, Washington, D.C. (2018)
5. Zhou, M., Qu, X., Jin, S.: On the impact of cooperative autonomous vehicles in improving freeway merging: a modified intelligent driver model based approach. IEEE Trans. Intell. Transp. Syst. **18**(6), 1422–1428 (2017)
6. Laval, J.A., Leclercq, L.: Microscopic modeling of the relaxation phenomenon using a macroscopic lane-changing model. Transp. Res. Part B: Methodol. **42**(6), 511–522 (2008)
7. Kober, J., Peters, J.: Imitation and reinforcement learning. IEEE Robot. Autom. Mag. **17**(2), 55–62 (2010)
8. Guo, J., Liu, Y., Zhang, L., Wang, Y.: Driving behaviour style study with a hybrid deep learning framework based on GPS data. Sustainability **10**(2351) (2018)
9. Mnih, V., Kavukcuoglu, K., Silver, D., Rusu, A.A., Veness, J., Bellemare, M.G., et al.: Human-level control through deep reinforcement learning. Nature **518**(7540), 529–533 (2015)
10. Sallab, A.E., Abdou, M., Perot, E., Yogamani, S.: Deep reinforcement learning framework for autonomous driving. Electron. Imaging **19**, 70–76 (2017)
11. Nagel, K., Schreckenberg, M.: A cellular automaton model for freeway traffic. J. De Phys. I **2**(12), 2221–2229 (1992)
12. Liu, Y., Guo, J., Taplin, J., Wang, Y.: Characteristic analysis of mixed traffic flow of regular and autonomous vehicles using cellular automata. J. Adv. Transp. (2017)
13. Sutton, R.S., Barto, A.G.: Reinforcement Learning: an Introduction. The MIT Press (1998)
14. Chen, D.J., Ahn, S., Chitturi, M., Noyce M.D.: Towards vehicle automation: roadway capacity formulation for traffic mixed with regular and automated vehicles. Transp. Res. Part B **100**, 196–221 (2017)
15. Ghiasia, A., Hussaina O., Qian Z., Li, X.: A mixed traffic capacity analysis and lane management model for connected automated vehicles: a markov chain method. Transp. Res. Part B **106**, 266–292 (2017)

Chapter 21
Data-Driven Disruption Response Planning for a Mass Rapid Transit System

Chunling Luo, Xinrong Li, Yuan Zhou, Aakil M. Caunhye, Umberto Alibrandi, Nazli Y. Aydin, Carlo Ratti, David Eckhoff and Iva Bojic

Abstract This paper studies the disruption management of a Mass Rapid Transit (MRT) network with data obtained from transportation smart cards. We introduce an optimization model for the development of efficient bus bridging services to minimize the negative effects of MRT disruption. Compared with existing approaches, our model considers the available capacity of existing buses when designing the routes and headways/frequencies of bus bridging services. The proposed model is demonstrated through one case study that assumes MRT disruption in the central business district area of Singapore. The case study shows that our approach can effectively reduce the travel delay of commuters and increase the number of commuters that can be served.

21.1 Introduction

For many of its early years, the Singapore Mass Rapid Transit (MRT) network system experienced disruptions only rarely, leading to it being recognized internationally for its efficiency and efficacy. According to a rail service reliability performance report from 2017, the number of these MRT disruptions, which resulted in service delays of

C. Luo (✉) · X. Li · I. Bojic
Singapore-MIT Alliance for Research and Technology, Singapore, Singapore
e-mail: chunling@smart.mit.edu; iseluoc@gmail.com

Y. Zhou · D. Eckhoff
TUMCREATE, Singapore, Singapore

A. M. Caunhye
The University of Edinburgh, Edinburgh, UK

U. Alibrandi
Aarhus University, Aarhus, Denmark

N. Y. Aydin
Singapore-ETH Centre, Singapore, Singapore

C. Ratti
Massachusetts Institute of Technology, Cambridge, MA, USA

© Springer Nature Singapore Pte Ltd. 2019
X. Qu et al. (eds.), *Smart Transportation Systems 2019*, Smart Innovation,
Systems and Technologies 149, https://doi.org/10.1007/978-981-13-8683-1_21

more than 30 min, has increased from 9 in 2011 to 16 in 2016 [8]. As the Singapore MRT network continues to be expanded with new trains and rails and the ridership continues to increase, the need for resilience improvement strategies becomes even more important.

Namely, the average daily ridership has grown from 2.3 in 2011 to almost 3 million passenger trips in 2016. Making the MRT network resilient to these disruptions is a challenging task that presents a vast array of research opportunities. By resilience we mean *the capacity of a system to absorb disturbance and to reorganize so as to retain essentially its structures, functions and feedback loops* [10]. Improving the resilience of the Singapore MRT network requires thus both proactive and reactive planning. Proactive planning concerns the addition of robustness, or the ability to absorb and mitigate shocks to the network. Reactive planning is themed around emergency response and is concerned with how the network responds to disruptions. Until now, resilience in the MRT has been studied at a more granular operational level, but recently it has been recommended that a more holistic perspective on system resilience should be taken [8].

A series of studies on network disruption management, including proactive and reactive planning, have been proposed in the literature to improve network resilience. For example, the disruption management process and the roles of different actors were discussed in [3]. This paper described three main challenges in disruption management: timetable adjustment, rolling stock and crew rescheduling. A summary of the algorithms developed for these three challenges for the Netherlands Railway company was provided by [7]. An integrated model for timetable and rolling stock rescheduling was developed by [1, 2], which minimizes the recovery time, the commuters inconvenience and system costs. Besides these studies on reactive planning, studies on proactive planning is also seen in the literature, such as an approach to enhance the resilience for MRT networks with localized integration with public bus services proposed in [4].

One of the most critical areas in network disruption management is the design of bridging services that can provide temporary transportation services in the disrupted parts to reduce the negative effects of disruption. Different bridging approaches have been adopted in practice, such as diverting disrupted commuters to other operating lines or other parallel public transport services, hiring taxi or bus bridging. A survey of disruption response management practices can be found in [9]. Among these response strategies, bus bridging services, which provide temporary bus transportation services to commuters, is the most common practice undertaken by transport operators in case of MRT disruption. Several approaches have been proposed specifically for bridging services in case of disruption. For example, an approach proposed in [11] suggests to examine whether and how to hire taxis to provide bridging services for short-term disruptions in public tram systems. Another paper proposed a comprehensive modeling framework and decision support system for planning and designing an efficient bus bridging network [6]. A different bus bridging service design was proposed by [5], where bridging bus routes were generated by a column generation procedure and the most effective combination of bus routes was identified via a path-based multi-commodity flow model.

However, these studies do not consider the integration between the bridging services and the existing bus routes. Namely, when MRT disruption occurs, the existing bus routes could provide complementary bus services to MRT network and divert some disrupted commuters to their destinations or other operating lines. Hence, whether and how to introduce bridging services should depend on the available complementary capacity of existing bus routes.

With the development of new technologies aimed at improving the information available about public transport systems, a large amount of data has become available in the process of building a resilient MRT network. With the help of data obtained with transportation smart cards (e.g., EZ-Link card in Singapore) and public MRT/bus line information, we are able to elicit the origin–destination demands and available complimentary capacity of existing buses, which can be helpful for the design of bus bridging services. In this paper, we thus propose an approach to develop efficient bus bridging services to minimize the negative effect of MRT disruption. Our model accounts for the available complementary capacities of existing bus routes when designing the routes and frequency of bridging buses. We demonstrate our methodology on one case study of MRT disruption in Singapore's Central Business District (CBD) area.

21.2 Methodology

We develop a mathematical optimization model that allows bus bridging services to be integrated with the existing bus services. Namely, when MRT disruption occurs, disrupted commuters can either make use of the complementary services provided by existing buses or services provided by bridging buses. The proposed model not only designs the bus bridging service routes, but also determines the frequency of each route.

21.2.1 In/Out Passenger Flows Dataset

The basis for our approach is comprehensive passenger data recorded for Singapore MRT and bus services for a duration of three months. Each record in this dataset has a timestamp of tap-in/out together with an MRT/bus stop identification. With that information, for each smart card, we can reconstruct the traveled route. Additionally, we used information about latitudes and longitudes of MRT/bus stops, official records on MRT/bus service and MRT/bus line information including operational starting time and ending time, traveling time and frequency of MRTs/buses. Based on the dataset and the MRT/bus line information, we elicit the following information: (1) the number of commuters traveling between each pair of origin–destination stations; (2) the time table and routes of MRT/bus services; (3) the travel time of the commuters; (4) the available capacity of each MRT and bus line. This information will be used as an input of our model.

21.2.2 Bridging Services Response Plan

The design of bridging services includes two main steps: (1) generating a candidate set of bridging routes and (2) route selection and bus allocation. One common practice for generating a candidate set of bridging routes is to replicate the MRT services by running buses parallel to the disrupted rail lines. Other approaches include using shortest path algorithm [6], a column generation procedure [5], or to generate candidate bus bridging routes considering more factors such as the pattern of commuter travel demand, travel time and transfer. In this paper, we focus on the second step (i.e., route selection and bus allocation) as the candidate set of routes is one of the inputs of our model and can be generated using one of the aforementioned existing approaches.

Before presenting our optimization model, we have to define the following sets: (1) $\mathcal{R}$ denotes the set of bus routes r; $\mathcal{R}^0$ and $\mathcal{R}^+$ denote the set of existing bus routes and candidate bridging bus routes, respectively; (2) $\mathcal{K}$ denotes the union of disjoint commuter group k that includes all commuters who go from origin station o_k to destination station s_k; (3) $\mathcal{R}^k$ is used to denote the set of bus routes that connect bus station o_k and destination s_k, that is, commuters in group k can and only can take bus routes in $\mathcal{R}^k$; (4) L_r denotes the set of edges/legs l for bus route r.

Our optimization model aims at minimizing the total travel time of commuters after disruption, including the riding time on the bus and the waiting time. First, we adopt the time–space network proposed by [5] to model the time dimension. For each group k, we discretize the whole time period into $\bar{u}$ periods associated with demand $d_{(k,u)}$ such that $\sum_u d_{(k,u)} = D_k$, where D_k is number of commuters in group k during the considered period. For each bus route r, we discretize the whole period into $\bar{v}$ service slots. The set of services slots for route r is denoted by $\mathcal{B}_r := \{(r, v), \forall v = 1, 2, ...\bar{v}\}$. Commuters in group k arrive at bus station o_k at time $\tilde{t}_{(k,u)}$, wait for bus on route $r \in \mathcal{R}^k$ to come, board and travel on bus for c_{kr} units of time. If the coming bus is full, then the commuters have to wait for the next bus. Let $w((k, u), (r, v))$ denote the waiting time of commuters in group (k, u) when bus slot (r, v) is taken. We have: $w_{((k,u),(r,v))} = \max\{0, t_{(k,(r,v))} - \tilde{t}_{(k,u)}\}$, where $t_{(k,(r,v))}$ is the time when bus service slot (r, v) arrives at the origin station o_k of group k. The commuter cannot take bus slot that arrives o_k before his arrival and it is reasonable to assume that commuter will not be willing to wait for a very long time (say, longer than a limit $\bar{w}$). Hence, we define set Ω that excludes those impossible combinations of $((k, u), (r, v))$: $\Omega = \{((k, u), (r, v)) : t_{(k,(r,v))} - \tilde{t}_{(k,u)} \geq 0, w((k, u), (r, v)) \leq \bar{w}\}$, where $\bar{w}$ is the limit of waiting time. We generate the matrix of $w_{((k,u),(r,v))}$ for all (k, u) and (r, v) based on the time–space network and use it as input coefficients for the model. Our optimization model will not only select the bridging bus routes, but also simultaneously determine the frequency/headway of the selected bus routes and the allocation of available bus resources among each route.

We define a discrete set of bus deployment plans: $\mathcal{P}_r := \{(r, h) : h \in I, \forall h_r^{min} \leq h \leq h_r^{max}\}$, where each plan (r, h) is characterized by route index r and the bus headway h; h_r^{min}, h_r^{max} denote the minimum and maximum allowed headways for the route r, respectively. Let $\mathcal{P}^+$ be the union of $\mathcal{P}_r$, $\forall r \in \mathcal{R}^+$. The bus deployment plans for existing route $r \in \mathcal{R}$ are already determined and their headway is h_r^0, $\forall r \in \mathcal{R}^0$. Let $\beta((r, h), (r, v))$ be a binary variable that equals 1 if the bus deployment plan $(r, h) \in P_r$ covers bus services plot (r, v), and 0 otherwise. The decision variables for our model are:

- $y(r, h) \in \{0, 1\} : \forall r \in R^+, (r, h) \in \mathcal{P}_r$. $y(r, h)$ takes 1 if bus deployment plan (r, h) is employed and 0 otherwise.
- $\chi_{((k,u),(r,v))} \geq 0$: the number of commuters in group (k, u) who take bus service slot $(r, v) \in \mathcal{B}_r$.
- $\eta(k, u)$: the number of commuters in group (k, u) who are unable to get on any bus by the waiting time limit $\bar{w}$.

Let c_k^0 denote the travel time of a single commuter in group $k \in \mathcal{K}$ when no disruption occurs. The bus bridging route selection and deployment problem can be formulated as:

$$\min \sum_{((k,u),(r,v))\in\Omega} (c_{kr} + w_{((k,u),(r,v))} - c_k^0)\chi_{((k,u),(r,v))} + \sum_{(k,u)} \theta_{(k,u)}\eta_{(k,u)} \tag{21.1}$$

$$s.t. \sum_{(r,v)\in B} \chi_{((k,u),(r,v))} + \eta_{(k,u)} = d_{(k,u)}, \ \forall(k, u) \tag{21.2}$$

$$\sum_{(k,u)} \gamma_{(k,(r,l))}\chi_{((k,u),(r,v))} \leq \beta_{((r,h_r^0),(r,v))}Q_{(r,v)}^0, \ \forall(r, v) \in \mathcal{B}_r, \forall l \in L_r, \forall r \in \mathcal{R}^0 \tag{21.3}$$

$$\sum_{(k,u)} \gamma_{(k,(r,l))}\chi_{((k,u),(r,v))} \leq \sum_{h|(r,h)\in\mathcal{P}_r} \beta_{((r,h),(r,v))}Qy_{(r,h)}, \ \forall(r, v) \in \mathcal{B}_r, \forall l \in L_r, \forall r \in \mathcal{R}^+ \tag{21.4}$$

$$\sum_{h|(r,h)\in\mathcal{P}_r} y_{(r,h)} \leq 1, \ \forall r \in \mathcal{R}^+ \tag{21.5}$$

$$\sum_{(r,h)\in\mathcal{P}^+} n_{(r,h)}y_{(r,h)} \leq A^{max} \tag{21.6}$$

$$\chi_{((k,u),(r,v))} = 0, \forall r \notin \mathcal{R}^k \tag{21.7}$$

$$y_{(r,h)} \in \{0, 1\}, \forall(r, h) \in \mathcal{P}^+; \tag{21.8}$$

$$\chi_{((k,u),(r,v))} \geq 0, \forall((k, u), (r, v)); \quad \eta_{(k,u)} \geq 0, \forall(k, u) \tag{21.9}$$

The objective function (21.1) minimizes: (1) the total increase in travel time for commuters taking buses and (2) the number of commuters who cannot board, weighted by a penalty parameter $\theta_{(k,u)}$. Constraint (21.2) guarantees that the total number of commuters who boarded plus the number of commuters who did not board equals the travel demand.

Let $\gamma((k, (r, l))$ be a binary variable that takes 1 if leg l is used by commuter group k when they take bus route r and otherwise 0. Constraints (21.3) and (21.4) ensure that on each leg $l \in L_r$ of each bus service slot (r, v), the number of commuters on the slot (r, v) does not exceed the available bus capacity $Q_{(r,v)}^0$ of existing bus routes and total capacity Q of introduced bridging bus routes, respectively. Constraint (21.5) restricts that at most one bus deployment plan can be employed on each route. Constraint (21.6) guarantees that the total number of buses additionally deployed should not exceed the bus resource capacity A^{max}, where $n_{(r,h)}$ is the number of buses required for route r with headway set as h. Constraint (21.7) ensures that commuters in group k only take bus route r that connects o_k and s_k (i.e., $r \in \mathcal{R}^k$). Constraints (21.8) and (21.9) define the domain of decision variables.

21.3 Case Study

In this section we demonstrate our methodology by studying one hypothetical disruption case in the Central Business District (CBD) area of Singapore during morning peak hours (i.e., 7:00 AM–9:00 AM). This region was chosen as it covers the central part of the Singapore MRT network, with a mixture of residential areas, business areas as well as commercial ones. The case study concerns a single-direction disruption of the purple MRT line, where the MRT links from station A to station D are disrupted (see Fig. 21.1). We assume that the disruption lasts for the whole peak hour period and that we need to assign bridging buses to cope with the morning peak hour demand. Historical data shows that about 7,400 commuters would be affected if this disruption would actually take place as described. Figure 21.1 shows the MRT network (represented by solid lines) and the existing bus lines (represented by dashed lines) in the disrupted area.

As mentioned before, the existing buses can provide complementary bus services to the affected commuters. In total, about 100 buses on these lines normally would pass the disrupted area during the affected hours, with an average free capacity per bus per link approximately equal to 74. In order to deal with the consequences of the assumed disruption, we first generate the candidate bus bridging routes via replicating the MRT services, considering all possible bus routes parallel to the purple MRT line

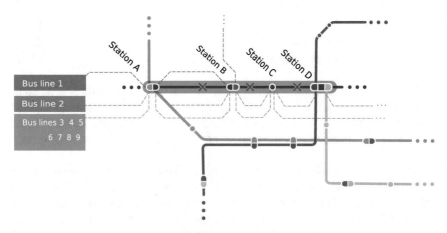

Fig. 21.1 Region we focus on. The affected MRT stations are: station A, B, C, D

Table 21.1 Optimal bus bridging services, where N stands for the number of allocated bus

Bridging bus route path	Headway (mins)	N
A → D	3	9
A → B → D	28.5	1

and then use the proposed optimization model to find the optimal route selection and bus allocation. Namely, the commuters can either be diverted to their destination if they alight in the disrupted area, or to station D to take other operating lines. Travel demand and available capacity of existing buses are derived from historical smart card data.

The parameters of the optimization model are set as follows: the maximum number of bridging bus A^{max} to be deployed is arbitrarily set to be 10 (later on we perform a sensitivity analysis to explore the impact of different bridging bus fleet sizes); minimum and maximum headways of bridging bus services are set to 1 min and 31 min respectively, with minimal incremental value set to be 30 s; the capacity of bridging bus Q is set to be 87 (the capacity of one type of bus in Singapore); the limit of waiting time $\bar{w}$ is set to equal 30 min and penalty parameter $\theta(k, u)$ is set to be 90 mins. The minimum and maximum headways are set considering factors such as headway ranges of existing buses, the high morning peak demand and the capacity of stations, while the penalty parameter is set to 90 min in order for it to be larger than the maximum waiting time (30 min) plus onboard time of disrupted commuters (shorter than 31 min).

The proposed model was coded in Python and solved in about 2 min by CPLEX V12.8.0 running on a personal computer with Intel Core i7 at 2.6 GHz and 16 GB RAM. The bridging services plan generated by our optimization model is shown in Table 21.1. As it can be seen, all of the bridging buses are allocated to divert commuters to station D from different affected MRT stations. This is because of two reasons: (1) all commuters who do not alight in the disrupted area are diverted to station D, which connects three MRT lines, to take other operating lines; and (2) only one of the existing bus lines (i.e., bus line 124 shown in Fig. 21.1) can provide complementary bus services to station D.

Furthermore, mainly due to resource constraints, other candidate routes are not selected, such as bridging routes that would directly connect stations C and D, probably as it could only provide a bridging service to one commuter group. Additionally, if the existing buses could provide enough complementary services for some commuter group, then there is no need to provide bridging bus for them. For example, B → C route is not selected because: (1) few commuters take MRT during morning peak hours for such a short distance; and (2) all of the existing bus lines except for bus line 1 can serve the affected commuters traveling that route.

Using the proposed bridging service plan, about 43.2 and 40.9% commuters can be served by the bridging buses and existing buses, respectively. We observe that the existing buses can serve up almost half of all affected commuters. Hence, in this case, the complementary capacities of existing buses are relatively important and should not be ignored when designing bridging bus plans.

Fig. 21.2 Percentage of
commuters served (higher is
better)

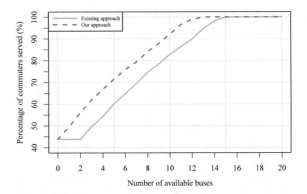

The red lines in Figs. 21.2 and 21.3 show the results of a sensitivity analysis performed to explore the impact of the bridging bus fleet sizes when using our approach. As the size increases, the average travel delay (including possible longer on-board time and waiting time for a bus) of all commuters decreases almost linearly and the percentage of served commuters increases. However, the average travel delay and percentage of served commuters almost no longer improve when the number of available buses exceeds 15, as all affected commuters are served. Assigning more than 15 buses would thus only increase costs, but would not generate any additional benefits. Moreover, we compare our approach with the existing approaches which do not consider complementary capacities of existing buses when allocating bridging buses.

The difference between the blue and red lines in Figs. 21.2 and 21.3 show that our approach always performs better in terms of both the average travel delay and the number of served commuters, especially when the number of available bridging buses is limited. In particular, when the number of available buses is less than 15, then our approach can serve about 10% more disrupted commuters than the existing approaches. This is mainly because existing approaches may allocate buses to serve commuters who can already be served by existing buses and result in oversupply on some routes and undersupply on other routes.

Fig. 21.3 Average travel
delay (lower is better)

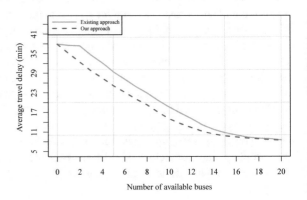

21.4   Conclusion

The optimization model proposed in this paper can be used for designing bus bridging services in response to MRT disruptions. Our approach can determine the routes to select and their corresponding headways in integration with the existing bus services, which can provide complementary services to MRT services in cases of disruption. We showed the effectiveness of our model via a case study in the central business district of Singapore. The results confirmed that our approach could generate solutions to effectively reduce the travel delay of commuters and the number of commuters who could not board a bus. In this paper, we considered only the cases when the travel demand and available capacities of existing buses are deterministic and derived from historical data. However, these parameters are actually subject to uncertainty. Hence, one potential direction of future work is to integrate uncertainty into our optimization model, which could be quantified by fitting a distribution to the data provided by the smart card records.

Acknowledgements   We acknowledge the National Research Foundation, Prime Minister's Office, Singapore under its Campus for Research Excellence and Technological Enterprise programme.

References

1. Cadarso, L., Marín, Á.: Recovery of disruptions in rapid transit networks with origin-destination demand. Procedia - Soc. Behav. Sci. **111**, 528–537 (2014)
2. Cadarso, L., Marín, Á., Maróti, G.: Recovery of disruptions in rapid transit networks. Transp. Res. Part E: Logist. Transp. Rev. **53**, 15–33 (2013)
3. Jespersen-Groth, J., Potthoff, D., Clausen, J., Huisman, D., Kroon, L., Maróti, G., Nielsen, M.N.: Disruption management in passenger railway transportation. In: Robust and Online Large-Scale Optimization, pp. 399–421. Springer (2009)
4. Jin, J.G., Tang, L.C., Sun, L., Lee, D.-H.: Enhancing metro network resilience via localized integration with bus services. Transp. Res. Part E: Logist. Transp. Rev. **63**, 17–30 (2014)
5. Jin, J.G., Teo, K.M., Odoni, A.R.: Optimizing bus bridging services in response to disruptions of urban transit rail networks. Transp. Sci. **50**(3), 790–804 (2015)
6. Kepaptsoglou, K., Karlaftis, M.G.: The bus bridging problem in metro operations: conceptual framework, models and algorithms. Public Transp. **1**(4), 275–297 (2009)
7. Kroon, L., Huisman, D.: Algorithmic support for railway disruption management. In: Transitions Towards Sustainable Mobility, pp. 193–210. Springer (2011)
8. LTA. Performance of rail reliability (2017)
9. Pender, B., Currie, G., Delbosc, A., Shiwakoti, N.: Planning for the unplanned: an international review of current approaches to service disruption management of railways. In: 35th Australian Transport Research Forum, Perth, Australia (2012)
10. Singapore-ETH Center. Future Resilient Systems programme booklet. (2014)
11. Zeng, A.Z., Durach, C.F., Fang, Y.: Collaboration decisions on disruption recovery service in urban public tram systems. Transp. Res. Part E: Logist. Transp. Rev. **48**(3), 578–590 (2012)

Chapter 22
Traditional Versus Budget Airlines—Comparison of Tickets Costs and Demands on the European Air Transportation Market

Marina Kholod, Yury Lyandau, Peter Golubtsov, Elena Okunkova and Nikolay Mrochkovskiy

Abstract This article discusses the market for passenger air travel, as well as the distinctive features of low-cost airlines and the specifics of their activities. The author draws attention to the significant growth and popularity of such companies, and asks the research question—is it really cheaper to fly budget airlines than traditional ones? To search for an empirically competent answer to the question posed, the article conducts a scientific and statistical study.

22.1 Introduction

In modern society, the aviation industry plays a tangible role in the global economy, especially when it comes to civil aviation. Traveling is an integral part of most people's lives, occupying an important place in it. According to statistics, almost 63 million people in the world work in this industry, and the impact of aviation on the world economy is estimated at about 2.7 trillion dollars a year, making it a key

M. Kholod (✉) · Y. Lyandau · P. Golubtsov
Chair of Management and Business Technologies, Plekhanov Russian University of Economics, 36 Stremyany Lane, 117997 Moscow, Russia
e-mail: kholod.mv@rea.ru

Y. Lyandau
e-mail: lyandau.yuv@rea.ru

P. Golubtsov
e-mail: golubtsov.pv@rea.ru

E. Okunkova
Chair of Organizational and Managerial Innovations, Plekhanov Russian University of Economics, 36 Stremyany Lane, 117997 Moscow, Russia
e-mail: okunkova.ea@rea.ru

N. Mrochkovskiy
Chair of Innovation Management and Social Entrepreneurship, Plekhanov Russian University of Economics, 36 Stremyany Lane, 117997 Moscow, Russia
e-mail: mrochkovskiy.ns@rea.ru

© Springer Nature Singapore Pte Ltd. 2019
X. Qu et al. (eds.), *Smart Transportation Systems 2019*, Smart Innovation, Systems and Technologies 149, https://doi.org/10.1007/978-981-13-8683-1_22

component of global business and one of the most popular industries in general. Every day, 8.6 million people use airline services. More than 99 thousand flights are carried out [1].

22.2 Theoretical Backgrounds

The rise in demand for the commercial air travel industry dates back to the 1960s; however, due to the high airline commission, ticket sales were not constant. This changed in the late 1970s, when the aviation industry was deregulated, and starting in the 1980s, civilian air travel successfully gained its popularity every year until September 11, 2001, when a terrorist attack took place in the United States of America before air travel. However, the influence of globalization and the expansion of the possibility of traveling without visas returned the demand of the population, and by the middle of the decade, demand for citizens in air travel began to grow again [2].

Of course, high demand for flights has also created a high level of competition between the airlines, thereby forcing them to find the most favorable conditions for the provision of their services [3]. Thus, budget airlines began to gain popularity, offering passengers to buy a ticket at the lowest price.

The term budget airline describes a company offering extremely low fare in exchange for the abandonment of most of the traditional passenger services, while the traditional airline is a standard airline with a set of free services already included in the price of the flight [4].

The first low-cost carrier appeared in the US in 1949—it was Pacific Southwest Airlines. However, the real boom in the development of this business model of airlines occurred in the late 1990s in Europe after the liberalization of the airline market. By the beginning of 2015, low-cost airlines took a leading position in the domestic air travel market of the EU [5].

The main difference between budget and traditional airlines lies in their business models. In order to reduce the price of a ticket for a potential buyer, a low-cost carrier tries to minimize all possible costs, thereby offering the cost solely for the flight itself, without including any additional services. Most often, such business model is characterized by the following features [6]:

- the use of one (relatively new) passenger class aircraft fleet, which saves on the training of personnel and technical maintenance of equipment;
- a minimum set of additional equipment in the aircraft, the lack of entertainment video panels, AVOD, ACARS pilot communication systems with the ground, auto braking, etc., which makes the cost of the car lower and reduces its weight, and hence the fuel consumption;
- policy of selling tickets online directly through airline websites

  - it allows to reduce the costs of paying commissions to agents and reservation systems;

- maximum load of the aircraft due to the provision of a larger number of seats, while minimizing the distance between the seats;
- use of secondary (less populated and, often, more distant from the city) airports to reduce airport charges, as well as a lower probability of delay (decrease in efficiency) of the aircraft;
- departures in the early morning and late evening to avoid delays due to congestion of airspace and lower airport charges;
- high aircraft turnover at airports and, as a result, the execution of more flights by one ship during the day (low-cost airlines planes make six flights a day, spending more than 10 h in the air, and only 30–60 min on the ground before each next flight—this time is usually enough for fueling and loading passengers);
- the existence of such routes, which are as popular and simple as possible, are mostly close in terms of distance and flight policy (the ability to save time on checking visa formalities). The route map is based on the "point-to-point" principle, that is, a direct flight without transit stops;
- the service and the number of services provided to the passenger free of charge are maximally reduced, which allows the company to receive a steady income for their sale (the cost of goods and services is also an order of magnitude higher than in traditional airlines);
- reducing the cost of maintaining a large staff (savings wage fund). Basically, the duties of an ordinary flight attendant also include additional functions (cabin cleaning, passenger check-in, checking baggage allowance, etc.);
- fuel cost hedging program;
- an increase in the cost of the ticket as the aircraft becomes full, which gives an advantage to the client (as well as the airline) only when booking the flight early.

Based on the foregoing, we can conclude that, unlike traditional airlines, low-cost airlines implement their business model to minimize costs and generate revenue for each additional service, which makes it possible on the one hand to offer the most attractive ticket price to a potential client, and on the other—get a substantial profit. This is what characterizes so high today the competitiveness of budget airlines.

Conducting an analysis of specialized and authoritative literature, as well as the official statistics of the International Air Transport Association (IATA), it can be noted that the phenomenon of low-cost airlines scored its comprehensive population rage is relatively recent, and its economic effect is characterized mainly by the following formula—"the difference of profits and costs multiplied by the number of flights". That is, the reduced ticket price for a potential client is compensated (for the airline) by an increased number of flights. However, if in a traditional airline the price of a flight may fluctuate in a relatively small range, then the price of a flight at the same low-cost carrier may vary by more than 10 times [7].

Today, the civil aviation industry covers almost any country, which, of course, makes this industry the subject of constant analysis in order to find its optimal use. The research question posed in this article is whether it's cheaper to fly a low-cost airline than traditional, will help determine whether the difference in financial gain is significant for the consumer who makes his choice in favor of budget companies [8].

To begin with, it is necessary to determine the market segment in which the scientific research will be conducted and its statistical sampling. When this condition is fulfilled, it is worth considering the fact that, depending on the distance of the flight, short-haul and long-haul flights are distinguished, respectively, referring to the previously described business model, it can be argued that long-haul flights are dominated by traditional airlines such as Aeroflot, Lufthansa or "American Airlines". This means that the analyzed market segment should be based on short-haul flights, including the service of both traditional airlines and budget ones [9].

The best example of such a segment is the European civil aviation market, which is one of the most sought-after and competitive in the world. Passenger traffic on the European market is more than 26.6% of the global, and the list of airlines serving it grows every year. Of course, in such tough competition, companies have to find all possible ways to maximize costs and increase their income. These factors influenced such a widespread and widespread distribution of budget airlines in Europe [10].

22.3 Data Collection and Data Analysis

According to the latest statistics, the top 15 most popular airlines in Europe (Table 22.1) include budget companies, however, they account for about 56% of the total passenger transportation from this list. It should also be noted that the average workload ratio of budget companies is higher than that of traditional (82.5% vs. 79.6%).

The table shows that in terms of demand budget airlines are not at all inferior to the traditional ones, but on the contrary, over time, they are gaining more and more popularity and distribution (Table 22.2) [6].

Table 22.2 shows how the market share of low-cost airlines has grown over the past 7 years, which, of course, characterizes the undeniable interest of the consumer in this type of service and services.

The financial statements also characterize the prospects and scale of growth of the budget model of airlines, because according to official data, the ultra-budget carrier Ryanair (Ireland) in 2015 produced the largest number of passenger traffic in the EU, receiving a net profit of 532 million euros (against 330 million euros for the German aviation giant Lufthansa).

Analyzing the above statistics, it becomes unclear why budget airlines get a greater net profit than traditional ones, despite the fact that the price of their ticket can be several times cheaper for a potential consumer. Consider this factor in more detail. Figure 22.1 shows the data on which specific areas there is a reduction in costs in budget airlines.

As can be seen from the diagram, low-cost airlines get a significant advantage over traditional ones due to a competent approach to cost reduction. According to

Table 22.1 15 European airlines by number of passengers transported for the year 2016 (million passengers)

	Company	Country	Type	2016	Passenger load, %
1	Ryanair	Ireland	Budget	106.4	82
2	Lufthansa	Germany	Traditional	74.7	78.1
3	EasyJet	UK	Budget	58.4	90.4
4	AirFrance	France	Traditional	50.6	81.5
5	British Airways	UK	Traditional	37.6	79.9
6	Air Berlin	Germany	Budget	33.3	80.7
7	Turkish Airlines	Turkey	Traditional	26.6	77.7
8	SAS scandinavian	Sweden	Traditional	25.5	75.1
9	KLM	Netherlands	Traditional	25.1	85.7
10	Alitalia	Italy	Traditional	24.3	74.6
11	Norwegian AS	Norway	Budget	17.7	78.5
12	SWISS IA	Switzerland	Traditional	15.8	82.9
13	Iberia	Spain	Traditional	14.8	81.5
14	Vueling Airlines	Spain	Budget	14.8	77.7
15	WizzAir	Hungary	Budget	12.4	85.7

Table 22.2 The market share of airlines in terms of passenger traffic in the European market for 2007–2014 (%)

Type	2007	2008	2009	2010	2011	2012	2013	2014
Budget	27.8	34.5	37.5	44.3	50.5	50	52	56
Traditional	72.2	65.5	62.5	55.7	49.5	50	48	44

the European Pilot Association, back in 2010, low-budget airlines saved 57% more on 11 items of expenditure compared with their competitors [9].

This means that, in comparison with traditional ones, budget airlines optimize their costs more successfully and often receive large profits. However, the question remains: do they really offer a potential ticket to a potential client several times cheaper? In order to find a competent answer to the research question posed, a scientific analysis was carried out, including the determination of a statistical sample of 15 traditional and 15 budget airlines (Table 22.3), and comparing the cost of their tickets to the same flight segment and the same date.

The analysis was carried out using the global airline search tool—the Skyscanner online server, which provides complete information on all possible flights of all airlines in the world. The variables were selected: a certain segment of the flight (the most popular European flight London–Paris), the price, the number of flights per day and the date of departure. To obtain the most detailed result, two analyses were carried out on one flight segment—with the date of departure several days later

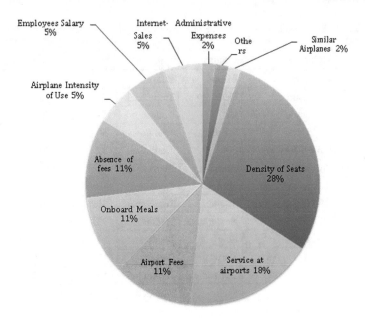

Fig. 22.1 Optimization of airline costs

Table 22.3 The largest and most popular airlines in Europe

Budget Airlines	Traditional Airlines
Ryanair	Air France
Easy Jet	British Airways
Norwegian	Swiss Int'l Air Lines
Vueling Airlines	Austrian
Wizz Air	KLM
Eurowings	Lufthansa
Volotea	Finnair
Air Europe	Aegean Airlines
Transavia Airlines	Iberia
AirBaltic	TAP Air Portugal
Smart wings	Alitalia
AirBerlin	Brussels Airlines
Jet2	Czech Airlines
Tuifly	SAS Airline
Condor	Adria Airways

Table 22.4   The cost of the ticket London–Paris a few days before departure

Budget Airlines	Price	#flights/day	Traditional Airlines	Price	#flights/day
Ryanair	65	2	Air France	116	1
Easy Jet	94	3	British Airways	163	2
Norwegian	255	1	Swiss Int'l Air Lines	304	1
Vueling Airlines	135	2	Austrian	218	1
Wizz Air	155	1	KLM	193	1
Eurowings	100	1	Lufthansa	240	1
Volotea	82	2	Finnair	230	1
Air Europe	220	1	Aegean Airlines	267	1
Transavia Airlines	79	2	Iberia	181	1
AirBaltic	–	–	TAP Air Portugal	–	–
Smart wings	174	1	Alitalia	190	1
Airberlin	216	1	Brussels Airlines	–	–
Jet2	–	–	Czech Airlines	180	1
Tuifly	–	–	SAS Airline	274	1
Condor	180	1	Adria Airways	–	–

(Table 22.4) and the date of departure one month from the day of the ticket search (Table 22.6).

The following is the descriptive statistics of the above data:

From Table 22.5 it can be seen that the minimum value of the compared airlines differs almost twice, while the maximum value is not so distinct, although the scope remains at the same level; dispersion also diverges significantly. With the same sample size, the average price of low-cost airlines and traditional ones differ relatively little—146 and 213 euros. However, the cost of a ticket from traditional carriers already includes airport check-in, transit, checked baggage, food, and drinks on board, as well as the possibility of accumulating bonus miles and a number of other advantages. Moreover, the cost of a budget carrier ticket includes only flights; other services can only be purchased additionally and for a high fee.

Analysis of the data showed that by purchasing an air ticket a few days before the departure date, the budget carrier does not always offer the potential consumer a favorable price, but most likely, the price will be based on the same level as that offered by traditional airlines.

Table 22.6 presents the data of the same airlines with prices for the flight in a month from the date of purchase of the ticket.

From Table 22.7 you can see that the minimum cost of the flight in the budget and traditional airlines differs by almost six times. This indicator is very high, while the maximum price value also differs by more than two times, as well as the variance. With the same sample size, the average price of low-cost airlines and traditional ones differ several times—88 and 215 euros, respectively. Even considering that additional

Table 22.5 Descriptive statistics of data

Descriptive statistics	Budget Airlines	Traditional Airlines
Sample size	12	12
Average value	146.25	213
Dispersion	4,033.11	2,810.18
Standard deviation	63.50	53.01
Standard error (average)	18.33	15.30
The coefficient of variation	0.43	0.24
Minimum	65	116
Maximum	255	304
Swipe	190	188
Median	145	205.5
Mean deviation	53.75	42.5
Amount	1,755	2,556
Sum of squares	301,03	575,34
Alternative asymmetry	0.314	0.026

Table 22.6 The cost of the ticket London–Paris a month before departure

Budget Airlines	Price	#Flights/day	Traditional Airlines	Price	#Flights/day
Ryanair	20	2	Air France	124	1
Easy Jet	86	3	British Airways	117	2
Norwegian	151	1	Swiss Int'l Air Lines	335	1
Vueling Airlines	30	2	Austrian	209	1
Wizz Air	65	1	KLM	213	1
Eurowings	132	1	Lufthansa	230	1
Volotea	93	2	Finnair	259	1
Air Europe	139	1	Aegean Airlines	278	1
Transavia Airlines	30	2	Iberia	171	1
AirBaltic	–	–	TAP Air Portugal	–	–
Smart wings	121	1	Alitalia	291	1
Airberlin	117	1	Brussels Airlines	–	–
Jet2	–	–	Czech Airlines	159	1
Tuifly	–	–	SAS Airline	188	1
Condor	67	1	Adria Airways	–	–

Table 22.7 Descriptive statistics of data

Descriptive statistics	Budget Airlines	Traditional Airlines
Sample size	12	12
Average value	87.58	214.5
Dispersion	2,071.35	4,548.09
Standard deviation	45.51	67.43
Standard error (average)	13.13	19.46
The coefficient of variation	0.51	0.31
Minimum	20	117
Maximum	151	335
Swipe	131	218
Median	89.5	211
Mean deviation	37.91	53.41
Amount	1,051	2,574
Sum of squares	114,83	602,15
Alternative asymmetry	−0.19589	0.21524

services from low-cost airlines are quite high, it will be much more profitable to purchase an air ticket from the financial side.

Thus, the research question—is it really cheaper to fly budget airlines than traditional ones—received an empirical justification. The difference in financial gain for the consumer who chooses budget airlines is significant only when purchasing a ticket for the flight in advance (approximately one month before the departure date). Analysis of the data showed that by purchasing an air ticket a few days before the departure date, the budget carrier does not always offer the potential consumer a favorable price, but, most likely, the price will be based on the same level as that offered by traditional airlines.

22.4 Conclusion

This study was conducted on the European aviation market for passenger transportation, because today the civil aviation industry (its diversity) in Russia is significantly lagging behind, which means the need for more detailed study and adoption of the experience of budget airlines in Europe. This will increase the competitiveness and relevance of the Russian civil aviation market and will make aviation services more affordable, high- quality and cheap for the consumer, and the industry itself will be bigger and more profitable, because the niche of low-cost air travel in Russia and the CIS is still free.

References

1. ATAG Report: Aviation benefits beyond borders. 1–4 July 2016
2. Annual Analyses of the EU Air Transport Market: Executive Summary (2015)
3. IATA Homepage: http://www.iata.org/Pages/default.aspx, Last accessed 09 Apr 2018
4. Report ECA: Upheaval in the European skies, pp. 19–20 (2010)
5. Rozenberg, R., Szabo, S., Sebescakova, I.: Comparison of FSC and LCC and their market share in aviation. Int. Rev. Aerosp. Eng. **7**(5), 149–154 (2014)
6. Diconu, L.: The evolution of the European low-cost airlines' business models. Ryanair case study. In: WC-BEM 2012, pp. 342–346 (2012)
7. Dobruszkes, F., Givoni, M., Vowles, T.: Hello major airports, goodbye regional airports? recent changes in European and US low-cost airline airport choice. J. Air Trans. Manag. **9**, 50–62 (2017)
8. De Wit, J., Zuiberg, J.: Route churn: an analysis of low-cost carrier route continuity in Europe. J. Transp. Geogr. **50**, 57–67 (2016)
9. Halpern, N., Graham, A., Dennis, N.: Low cost carriers and the changing fortunes of airports in the UK. Res. Trans. Bus. Manag. **21**, 33–43 (2016)
10. Tomova, A., Ramajova, L.: Frequent flyer programs and low-cost airlines: ongoing hybridization?. Procedia Soc. Behav. Sci. **110**, 787–795 (2014)

Chapter 23
Blockchain Applications in Shipping, Transportation, Logistics, and Supply Chain

Shuaian Wang and Xiaobo Qu

Abstract Blockchain, as a new technology, has aroused wide attention from industry and academia in recent years. There are a huge number of articles and news reports about blockchain applications in shipping, transportation, logistics, and supply chain. This short tutorial aims to clarify (i) the key characteristics of blockchain that enable its use in shipping, transportation, logistics, and supply chain and (ii) how these key characteristics are applied. The key characteristics are decentralization, open data, and authenticity of data. The applications include smart contract, fast payment, information sharing, track and trace, and supply chain finance. We hope this tutorial can provide introductory knowledge to practitioners in shipping, transportation, logistics, and supply chain people who are not familiar with blockchain.

23.1 Introduction

Most people have heard the buzz words "bitcoin" and "blockchain". Bitcoin is one of the earliest, and perhaps the most important to date, application of blockchain. Nowadays, the production and distribution network is globalized and of high complexity, which requires new innovations to address the challenges therein. Blockchain is one of such new innovations. There have been a number of blockchain applications in shipping, transportation, logistics, and supply chain. There were 89 million Google pages when searching "blockchain AND shipping OR transportation OR logistics OR supply chain" on December 25, 2018.

Blockchain is a distributed digital ledger system. First, blockchain is a ledger system, a system that records the transactions. It should be noted that here transactions

S. Wang
Department of Logistics and Maritime Studies, Hong Kong Polytechnic University, Hung Hom, Hong Kong
e-mail: wangshuaian@gmail.com

X. Qu (✉)
Department of Architecture and Civil Engineering, Chalmers University of Technology, 41296 Gothenburg, Sweden
e-mail: xiaobo@chalmers.se

© Springer Nature Singapore Pte Ltd. 2019
X. Qu et al. (eds.), *Smart Transportation Systems 2019*, Smart Innovation,
Systems and Technologies 149, https://doi.org/10.1007/978-981-13-8683-1_23

do not necessarily involve money. A transaction in blockchain should be understood as a new piece of information, e.g., an arrival of truck to a warehouse and a recording of inner temperature of a refrigerated container. Second, blockchain is a digital system. This is easy to understand: blockchain involves a huge number of transactions and many players, and a paper-based system cannot work. Third, blockchain is a distributed system. There are many players but there is no central agency that controls or coordinates the players, such as a bank.

Blockchain can be considered as a WhatsApp (or WeChat) group, where (i) all relevant players are in the group, (ii) each player will post in the group all relevant transaction information, (iii) all players will download all the chat records to local disk, and Facebook (or Tencent) does not interfere with the chat in the group. By likening blockchain to a WhatsApp group, we can understand the following three key characteristics that enable blockchain to be applied in shipping, transportation, logistics, and supply chain.

- Decentralization: All the players can communicate directly with each other (note that we have assumed Facebook (or Tencent) does not interfere).
- Open data: All the transaction information is in the chat history of the group, and all the players can access the information.
- Authenticity of data: No player can modify his historical chat records or fake the other players' chat records.

23.2   Overview of Blockchain Applications

We summarize the blockchain applications in shipping, transportation, logistics, and supply chain in Table 23.1. The applications are classified into several categories, including smart contract, fast payment, information sharing, track and trace, and loan from a bank. We provide a few examples for each category and these examples will be explained in subsequent sections. We also list the techniques that enable each application, including workflow management, Internet of Things (IoT), the decentralization (D) characteristic of blockchain, the open data (O) characteristic, and the authenticity of data (A) characteristic.

23.3   Smart Contract: Workflow Management Instead of Blockchain

DHL reported that 10% of freight invoices contain errors and a reduction of 5% transport cost reduction can be achieved by eliminating the errors [1]. Winnesota estimated that $140 billion is associated with disputes in payment in the transport industry [2]. Winnesota further stated that 20% of transportation costs are related to reliance on paperwork [2]. Smart contract can significantly reduce bureaucracy

Table 23.1 Overview of blockchain applications

Application	Examples	Workflow	IoT	Decentralization	Open data	Authenticity of data
Smart contract	Automatic delivery	✓				
Fast payment	Bitcoin			✓		
Information sharing	US truck companies				✓	
Track: transport stage	Maersk Line	✓	✓			
Track: environment	Cold chain		✓			✓
Track: process	Odometer		✓			✓
Trace: authenticity	Luxury products					✓
Trace: illegal	Ivory products					✓
Trace: unethical	Diamond					✓
Trace: unsustainable	Plastics					✓
Trace: quality management	Walmart					✓
Supply chain finance	Loan from bank					✓

and paperwork. An example of a smart contract is that once payment is received, the delivery of product is automatically triggered. Another example is that once the freight is received, payment is automatically triggered. A related application is that shipping company ZIM digitalizes the bill of lading in its trial of a blockchain-based system.

Smart contract is often quoted as an advantage of blockchain. However, smart contract is not the innovation of blockchain. Smart contract is part of workflow management, which appeared much earlier than blockchain. Although smart contract can be integrated into blockchain, it is not a characteristic of blockchain.

23.4 Fast Payment

Winnesota estimated that on average it takes a company 42 days to receive payment [2]. Using blockchain, payment can be done much faster, say, in 1 h. Australian car maker Tomcar uses bitcoin to pay some of its suppliers. The fast payment is enabled by the decentralization characteristic of blockchain, eliminating the central agency that is slow. (It should be noted that there are disadvantaged of payment using bitcoin, which will not be discussed in the paper.)

23.5 Information Sharing

DHL estimated that there are over 500,000 trucking companies in the US [1]. Winnesota estimated that 90% of the trucking companies have at most six trucks [2]. Because of spatial and temporal transport demand unbalance, these companies have low truck utilization in less-than-container-load (LCL) transport. Winnesota estimated that trucks travel 29 billion miles per year with partial or empty loads [2].

Blockchain in Transport Alliance (BiTA) is an organization that aims to improve trucking efficiency using blockchain. By sharing truck and cargo information, trucking companies can reduce empty trips and decrease their costs. This application is enabled by the open data characteristic of blockchain. It should be noted that a third-party platform may also function in a similar way. The disadvantages of a third-party platform, compared with blockchain, include: the former may seek to make a profit and hence distort the market (e.g., by prioritizing the trucking companies that pay more to the platform) and the former may be the target of cyberattack while blockchain has much higher cybersecurity.

23.6 Track and Trace

The track and trace functions are arguably the most important applications of blockchain. "Track" means monitoring the process and "trace" means uncover the origin. Track and trace utilize the authenticity of data characteristic of blockchain. The tracking applications and the tracing applications are elaborated below.

23.6.1 Track

Maersk Line established a global blockchain-based system to track container shipment. Similar to smart contract, this type of tracking application is not the innovation of blockchain, either. It should be noted that the tracking of transport stages may need other technologies in IoT, such as radio frequency identification (RFID) tags.

A very important application is tracking the environment while a product (e.g., food, flower, medicine) is in transit. Winnesota estimated that 8.5% of sensitive pharmaceutical products are discarded due to exposure to unacceptable temperatures in transit [2]. Blockchain can integrate IoT technologies, that is, sensors measuring temperature, humidity, vibration, and other items of interest. The recordings of the sensors can be stored on blockchain and hence are tamper free and open to all players.

Fig. 23.1   A refrigerator in Korea Market, New Zealand

Figure 23.1 shows a refrigerator of products and the display of the temperature. If the sensors for temperature are operated by a third party and the temperature is regularly stored on a blockchain, then customers can be confident that the products have been stored in appropriate conditions. Depending on the deviation from the required range of parameters of the environment, actions triggered by the deviation may include adjustment of the environment settings, changing the expiry date of the product, declaring the product is unfit, and applying penalties to the player that is in charge of ensuring an agreed environment.

The third application of blockchain is to track the process to make the tampering of results impossible or at least more costly. DHL estimated the odometers of one-third of second-hand cars in Germany have been illegal manipulated [1]. Bosch and TÜV Rheinland created a blockchain-based system with an in-car sensor to regularly record the mileage of each vehicle. The recorded data are stored on a blockchain.

23.6.2   Trace

Tracing the origin of a product, or the origins of the raw materials, components, and sub-assemblies of a product, brings in a number of benefits. (i) Tracing the origin can ensure the authenticity of a product, especially for expensive products such as diamond, handbags, and pharmaceutical products. (ii) Tracing the origin can ensure that a product is legally manufactured. For instance, the trade of elephant ivory is generally illegal but the trade of the ivory from the tusks of dead woolly mammoths frozen in the tundra is legal. (iii) Tracing the origin can ensure ethical activities. For example, De Beers uses blockchain technology to track stones from the point they are mined all the way to the point when they are sold to consumers to avoid "blood diamonds". (iv) Tracing the origin can promote the use of sustainable products. Figure 23.2 shows a bottle of laundry liquid in a supermarket with the tag "All our bottles are made from sugar plastics! 100% kerbside recyclable". If the origin of the bottles is authentic and stored on a blockchain, then environment-conscious customers will buy more of this brand of laundry liquid. (v) Tracing the origin can assist quality management. Walmart uses blockchain to track sales of pork meat and checks where each piece of meat comes from, its processing and storage, and its sale to customers.

Fig. 23.2 A bottle of laundry liquid in a PAK'nSAVE supermarket, New Zealand

23.7   Supply Chain Finance

Many supermarkets adopt the vendor-managed inventory (VMI) policy, that is, the products belong to the supplier of the supermarkets before they are sold. With blockchain, banks can be sure that the products indeed belong to the supplier, and hence the products can serve as a warrant for the supplier to take a loan from banks. Logistics transactions are often paid monthly or quarterly. Therefore, a company may have a huge amount of money to receive. With blockchain, the money to be received can be guaranteed, and this facilitates the company to borrow money.

References

1. DHL: Blockchain   in   logistics.   https://www.logistics.dhl/content/dam/dhl/global/core/documents/pdf/glo-core-blockchain-trend-report.pdf, Accessed Dec 2018
2. Winnesota: How blockchain is revolutionizing the world of transportation and logistics. https://www.winnesota.com/blockchain, Accessed Dec 2018

Chapter 24
Ship Inspection by Port State Control—Review of Current Research

Ran Yan and Shuaian Wang

Abstract Port state control (PSC) is an international regime to inspect the foreign ships coming to the port state in order to ensure that they are compliant with various international conventions. In this study, we conduct a review of the literature related to the PSC inspection in four areas: factors influencing the PSC inspection results, inspected ship selection scheme, effect of PSC inspection, and suggestions to improve PSC inspection. We found that both ship factors and non-ship factors would influence the PSC inspection outcomes, and the PSC inspection could improve the safety level of maritime industry and protect marine environment. Meanwhile, there is still room to improve the PSC inspection, including adopting more efficient methods to select high-risk ships, constructing combined databases and harmonization the PSC inspection authorities. It is expected that more research in improving and summarizing the PSC inspection will emerge in the coming years. We believe that this review can shed light on the development of PSC-related studies.

24.1 Introduction

Maritime transportation is playing an important role in the development of globalization [1]. Meanwhile, accidents and incidents aroused by maritime transportation render threats and risks to the marine environment [2]. To improve the situation, there are an increasing number of international conventions to maintain marine safety, crew training and pollution prevention. These regulations include but not limited to SOLAS, MARPOL, STCW, Tonnage Measurement, and Load Line conventions [3]. Generally, the flag state of the ships is responsible to ensure the ships flying its flag to meet the convention standards. However, many ships may irregularly visit their flag state ports and this would restrict the enforcement of flag state on its ships. Under

R. Yan · S. Wang (✉)
Department of Logistics and Maritime Studies, Hong Kong Polytechnic University, Hung Hom, Hong Kong
e-mail: wangshuaian@gmail.com

R. Yan
e-mail: angel-ran.yan@connect.polyu.hk

© Springer Nature Singapore Pte Ltd. 2019
X. Qu et al. (eds.), *Smart Transportation Systems 2019*, Smart Innovation,
Systems and Technologies 149, https://doi.org/10.1007/978-981-13-8683-1_24

such circumstance, the first Memorandum of Understanding (MoU) on Port State Control (PSC) was found in Europe in 1982, which is often referred to as "Paris MoU" [4]. The basic idea of PSC is that the port states can have the right to inspect the coming foreign ships to ensure that they are safe and unlikely to pollute their water [5]. With the support of the International Maritime Organization (IMO) and the International Labour Organization (ILO), the scale of Paris MoU is growing and the number of MoUs in the whole world is increasing. Nowadays, there are totally nine PSC MoUs all over the world aiming to eliminate substandard shipping.

Due to the limited resources and the high inspection cost, the PSC MoUs are unable to inspect all coming ships. As a result, one critical issue faced by the PSC MoUs is how to select ships to be inspected. Currently, one typical method adopted by PSC authorities is to set an inspection rate first in order to ensure the minimal number of inspected ships, then to use target factors attached with weighting points to identify high-risk ships. The target factors include ship characters (e.g., ship age, ship flag, ship recognized organization and ship type) and previous inspection records (e.g., last inspection time and precious detention time). In addition, the overriding factors, such as collision, grounding or stranding on the way to the port and an alleged pollution violation, etc., will also lead to an inspection [6].

When a ship is selected to be inspected, the port state control officer (PSCO) will first conduct an initial inspection, which mainly includes the first impression of the ship, certificate check and "walk around" to check the overall ship conditions. If major problems are found, clear grounds will be conducted before moving to an in-depth inspection, which is focused on the ship's equipment, construction, manning, and working and living conditions. When the inspection is finished, an inspection report will be generated, in which the conditions that do not obey the conventions are denoted as "deficiencies" (including the deficiency number and deficiency types). If the PSCO decides that the ship is unsafe to the maritime environment, the ship can be detained until the deficiencies are rectified. The deficiencies and detention of the ship are called the PSC inspection results. The inspection results combined with the ship information will be recorded in the database of the corresponding MoUs.

24.2 Literature Search Method and Review Structure

Since the first PSC program was introduced by the IMO in 1982, the PSC inspection has been receiving increasing attention from researchers and policy-makers. To access related papers, we searched the database of Google Scholar, Web of Science, and Scopus by using the keywords Port State Control or Port State Controls. A total of 43 papers that are principally focused on PSC inspection were found. In terms of their topics, we divide the literature into four areas: factors influencing the PSC inspection results, inspected ship selection scheme, effects of PSC inspection and ways to improve the PSC inspection. Specific, factors that influence the PSC inspection including ship factors and non-ship factors and the effects of PSC inspection contain the effects on maritime safety, effects on inspected ships and effects on

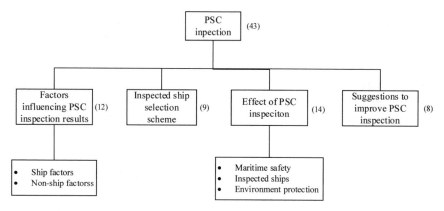

Fig. 24.1 Structure of the review

the environment protection. The overview of literature classification is illustrated in Fig. 24.1.

24.3   Factors Influencing the Results of PSC Inspection

Much of the current literature considers the factors that could affect the PSC inspection results. Some of those factors are ship factors, including the ship generic factors (e.g., ship age, ship type, ship size, the performance of ship flag and ship company, etc.) and ship inspection factors (e.g., the number of previous detentions, the number of outstanding deficiencies, etc.). A small number of studies focus on the non-ship factors, including the impact of PSC inspection time, inspection area and the background of the PSC inspectors. In some papers, the abovementioned factors have been analyzed simultaneously.

24.3.1   Ship Factors Influencing PSC Inspection Results

Some papers are focused on generic factors. Cariou et al. reported that ship age, type, and flag are the dominant predictors of ship deficiencies by using Poisson models constructed from 4,080 observations [7]. Cariou et al. then further identified that the determinants of PSC inspection results were ship age (40%), ship recognized organization (31%) and place of inspection (17%) based on 26,515 PSC inspections [8]. Recently, Yang et al. identified the factors that were most influential to bulk carriers' detention were the detected deficiency number, inspection type, ship recognized organization, and ship age by using a data-driven Bayesian network [9].

More detailed ship factors are also identified in the literature. Cariou presented a quantile regression model and concluded that bulk vessels, dry cargos, and reefer ships, as well as older ships were with a higher number of deficiencies and probability of detention [10]. Tsou used big data methods to analyze the inspection data and claimed that ships that are more than 25 years old, with less than 8,500 gross tonnages, of general cargo/multipurpose type, at some certain year and place to be inspected and of some certain flags might perform worse in the PSC inspections [11].

24.3.2 Non-ship Factors Influencing PSC Inspection Results

Regarding mainly to non-ship factors, Knapp and Franses were the pioneers who used the econometric methods to analyze the influencing factors. In 2007, they proposed a logistic regression model based on 18,319 inspections globally to identify the differences in probability of detention and identified deficiencies. They claimed that the inspection areas and different background of inspectors would influence the inspection results [12]. Ravira and Piniella, and Graziano et al. both pointed out that the professional profile of PSC inspectors might have an impact on the inspection results [13, 14].

The abovementioned papers all use statistical models (regression model, count data model, and variance decomposition analysis) to apply to case data sets and find out the determinant factors of PSC inspection results. The conclusions indicate that both the factors related directly to ships and to PSC inspections will both impact the inspection results.

24.4 Ship Selection Scheme in PSC Inspection

How to select inspected ships in the PSC inspection is a critical issue. Not all foreign ships coming to the port states will be inspected due to the limited time, budget and human resources. However, if substandard ships are not identified and the deficiencies are not rectified, the safety and health of maritime environment will be damaged.

A considerable amount of literature has been focused on the ship selection scheme to make the selecting process more efficient. For example, in 1999, Li proposed a new ship assessment system to automatically give each coming ship a risk score based on its age, flag, insurers, classifications and operators [15]. Similarly, Degré demonstrated a high-risk vessels selection scheme based on "Risk Concept" which combined ship generic variables with the Paris MoU criteria [16].

Several studies use the machine learning models for ship selection in PSC inspections. Xu et al. demonstrated a Support Vector Machine (SVM) risk assessment system based on target factors in order to classify the ships into high-risk group or low-risk group [17]. They then combined website scrapper technology to extract more target factors and then included in the SVM model to improve its accuracy

[18]. Based on these studies, Gao et al. further improved the classification accuracy of the proposed SVM model by combing K-Nearest Neighbor (KNN) method to remove noisy training examples [19]. In addition, Zhou and Sun introduced a new mathematical model which could be automatically optimized and self-evolved by using Generalized Additive Modeling (GAM) [20].

Overall, these studies all take factors related to ship itself, including the ship age, flag, type, company, and size into account when analyzing the results of PSC inspection. Conversely, the factors related to historical inspection information, such as the last inspection time, the previous number of detentions and the last inspection authority are seldom included in the above studies. In addition, the dynamic factors of the ships, such as the change of flag, change of ship company or classification society, and change of captain and sailors are not considered in the above papers. Regarding the methodology proposed in these studies, they all use mathematical models to quantitatively illustrate the influence of the factors on PSC inspection results. Compared with the ship selection methods adopted by most of the PSC MoUs, these models can identify substandard ships more efficiently and accurately.

24.5   Effect of PSC Inspection

While the flag state is seen as the first line of defense in eliminating substandard ships, the port state is seen as the second line [12]. After the first PSC program was introduced in 1982, a large volume of studies has discussed the effects of it. Basically, the effects of PSC inspection are in three aspects: effect on maritime safety, effect on inspected ships and effect on environment protection.

24.5.1   PSC Inspection Effect on Maritime Safety

The main research stream of this topic focuses on the effectiveness of PSC inspection on improving maritime safety. Knapp and Franses used combined data sources to identify how PSC inspections would affect the probability of casualty. They figured out that the casualty probability can be reduced after the PSC inspections conducted [12]. In addition, Li and Zheng concluded that the PSC programs were powerful in improving the maritime safety level by reducing total accident loss number and loss rate [5]. More specific, Knapp et al. provided a monetary quantification to estimate the incident cost savings brought by PSC inspection. They figured out that the estimated range of monetary benefit of PSC inspection was from about 70,000–190,000 dollars [21]. Regarding the specific relationship between ship inspection factors and accident involvement, in 2014, Hänninen and Kujala proposed a Bayesian network model and identified that the ship type, PSC inspection type and the number of deficiencies related to structural conditions were the most influential factors of accident involvement [22]. Hänninen et al. then constructed another Bayesian network model and claimed that ship safety management could be highly improved after the PSC inspection [23]. Recently, Heij and Knapp investigated the predictive power of the PSC inspection by identifying that a worse PSC inspection outcome in the previous year, the higher probability of shipping accident in the next year [24]. The above papers all aim to validate the effectiveness of PSC inspection. Regarding the effectiveness of PSC inspection on the Maritime Labor Convention, Grbić et al. stated that it was becoming forceful in detecting unacceptable working and living conditions for crew onboard [25]. On the contrary, Bateman pointed out that the PSC inspection appeared to be inefficient in reducing the substandard ships in the Indian Ocean Region (IOR) where a large number of piracy, robbery, and other illegal activities still existed [26].

Overall, although PSC inspection may be inefficient in some developing world due to the complicated conditions and limited resources, the introduction of PSC can help improve the safety of the global maritime environment, especially by reducing the occurrence of maritime accident and the maritime risk loss rate.

24.5.2 PSC Inspection Effect on Inspected Ships

Another research stream is the effect of PSC inspections on the inspected ships. Cariou et al. suggested that the deficiencies detected in the next PSC inspection would be 63% fewer than that of the last inspection [27]. As the ship selection schemes adopted by PSC MoUs took ship flag and classification society into account, Cariou and Wolff proposed a bivariate Probit model and figured out that the PSC inspection might lead to ships' decisions on changing flag (flag-hopping) and classification society (class-hopping) [28]. Fan et al. also argued that PSC inspection might increase the possibility of flag-out [29].

24.5.3 PSC Inspection Effect on Environment Protection

Apart from the abovementioned two research streams, some researchers also observed the influence of PSC inspection on environment protection. Titz pointed out that PSC inspection was confirmed to be effective in reducing marine environment pollution [30].

Together, these studies outline that PSC inspection is impactful in reducing maritime risks, improving the condition of ships and protecting the maritime environment. However, there may be some drawbacks brought by the PSC inspections. As the ship selection scheme for PSC inspection takes the performance of ship flag and classification society into account, this may give rise to opportunistic behaviors including ships' flag-hopping and class-hopping to reduce their inspection frequency.

24.5.4   Suggestions for PSC Inspection

A number of authors have given advice on PSC inspections in order to improve the inspection efficiency and accuracy. Bang and Jang identified the performance of nine MoUs were varied and thus advised establishing a system in which more advanced MoUs could assist those less advanced MoUs [31]. After interviewing experts, doing Analytic Hierarchy Process (AHP) analysis and evaluating questionnaires, Liou et al. suggested an optimal solution for restructuring Taiwan's PSC inspection authority was to establish an independent government agency under the Ministry of Transportation and Communications [32].

Harmonization of PSC MoUs' databases and combining with other inspection reports and casualty databases have been suggested in many papers, including but not limited to Knapp and Franses [33], Knapp and van de Veldon [34], Knapp and Franses [35] and Heij et al. [36].

Game models are also adopted in some papers. Li and Tapiero outlined a game model between the port authority and ship operators based on a random payoffs game–theoretical framework [37]. Yang et al. proposed a risk-based game model between shipowners and port authorities to figure out the optimal inspection policy. Based on the Nash equilibrium, port authorities optimal inspection rates and ship owners optimal maintenance rates were generated [9].

In view of all that has been mentioned so far, there is no doubt that PSC inspection is effective in rectifying substandard ships and improving maritime safety. Nevertheless, there is still room to develop its inspection strategies by adjustment of PSC inspection authority, combining databases of different MoUs, and with accident and casualty reports, or adopting some mathematical models to better trade-off the inspection costs and rates.

24.6   Future Research Opportunities

As the economic development and world trade depend largely on maritime transportation, PSC inspection will become a more and more essential guard of maritime safety. Undoubtedly, more studies will further evaluate as well as improve the effectiveness of PSC inspection. We will outline some promising future research directions of PSC inspection.

To develop more efficient ship selection schemes, one possible way is to combine different databases, including among different PSC MoUs and with the ship information databases, maritime accident and incident databases and databases of other inspections in order to get more comprehensive case data sets. Also, more historical data can be taken into account when developing ship selection schemes despite the access difficulties, as the ship inspection history is a strong indicator for future inspection results. In addition, more advanced methods can be adopted, especially those machine learning methods that are good at classification and prediction.

In recent years, marine environment protection and human factors of ship operations are receiving more attention than before [38, 39]. As a result, PSC authorities are focusing more on onboard living and working conditions as well as maritime pollution caused by substandard ships [36]. However, there are few papers that are related to the effect of PSC inspections on protecting the marine environment and improving onboard conditions. Thus, future research should further evaluate the effects of PSC inspections on these two areas.

References

1. UNCTAD.: Review of maritime transportation 2018. Accessed November 2018 from https://unctad.org/en/PublicationsLibrary/tdr2018_en.pdf
2. Yang, Z., Yang, Z., Yin, J.: Realising advanced risk-based port state control inspection using data-driven Bayesian networks. Transp. Res. Part A **110**, 38–56 (2018)
3. Veiga, J.L.: Safety culture in shipping. WMU J. Marit. Aff. **1**(1), 17–31 (2002)
4. Paris MoU.: Paris MoU annual report 2017 "Safeguarding responsible and sustainable shipping". Accessed December 2018 from https://www.parismou.org/2017-paris-mou-annual-report%E2%80%9Csafeguarding-responsible-and-sustainable-shipping%E2%80%9D
5. Li, K.X., Zheng, H.: Enforcement of law by the port state control (PSC). Marit. Policy Manag. **35**(1), 61–71 (2008)
6. Intercargo.: Port state control. A guide for ships involved in the dry bulk trades. Accessed December 2018 from https://www.mardep.gov.hk/en/others/pdf/pscguide.pdf
7. Cariou, P., Mejia Jr., M.Q., Wolff, F.C.: An econometric analysis of deficiencies noted in port state control inspections. Marit. Policy Manag. **34**(3), 243–258 (2007)
8. Cariou, P., Mejia, M.Q., Wolff, F.C.: Evidence on target factors used for port state control inspections. Mar. Policy **33**(5), 847–859 (2009)
9. Yang, Z., Yang, Z., Yin, J., Qu, Z.: A risk-based game model for rational inspections in port state control. Transp. Res. Part E **118**, 477–495 (2018)
10. Cariou, P., Wolff, F.C.: Identifying substandard vessels through port state control inspections: a new methodology for concentrated inspection campaigns. Mar. Policy **60**, 27–39 (2015)
11. Tsou, M.C.: Big data analysis of port state control ship detention database. J. Mar. Eng. Technol. 1–9 (2018)
12. Knapp, S., Franses, P.H.: A global view on port state control: econometric analysis of the differences across port state control regimes. Marit. Policy Manag. **34**(5), 453–482 (2007)
13. Ravira, F.J., Piniella, F.: Evaluating the impact of PSC inspectors' professional profile: a case study of the Spanish Maritime Administration. WMU J. Marit. Aff. **15**(2), 221–236 (2016)
14. Graziano, A., Cariou, P., Wolff, F.C., Mejia, M.Q., Schröder-Hinrichs, J.U.: Port state control inspections in the European Union: do inspector's number and background matter? Mar. Policy **88**, 230–241 (2018)
15. Li, K.X.: The safety and quality of open registers and a new approach for classifying risky ships. Transp. Res. Part E Logist. Transp. Rev. **35**(2), 135–143 (1999)
16. Degré, T.: The use of risk concept to characterize and select high risk vessels for ship inspections. WMU J. Marit. Aff. **6**(1), 37–49 (2007)
17. Xu, R.F., Lu, Q., Li, W.J., Li, K.X., Zheng, H.S.: A risk assessment system for improving port state control inspection. IEEE International Conference on Machine Learning and Cybernetics, vol. 2, pp. 818–823 (2007)
18. Xu, R., Lu, Q., Li, K. X., Li, W.: Web mining for improving risk assessment in port state control inspection. IEEE International Conference on Natural Language Processing and Knowledge Engineering, pp. 427–434 (2007)

19. Gao, Z., Lu, G., Liu, M., Cui, M.: A novel risk assessment system for port state control inspection. IEEE International Conference on Intelligence and Security Informatics (ISI), pp. 242–244 (2008)
20. Zhou, C., Sun, J.: Automatically optimized and self-evolutional ship targeting system for port state control. IEEE International Conference on Systems Man and Cybernetics (SMC), pp. 791–795 (2010)
21. Knapp, S., Bijwaard, G., Heij, C.: Estimated incident cost savings in shipping due to inspections. Accid. Anal. Prev. **43**(4), 1532–1539 (2011)
22. Hänninen, M., Banda, O.A.V., Kujala, P.: Bayesian network model of maritime safety management. Expert Syst. Appl. **41**(17), 7837–7846 (2014)
23. Hänninen, M., Kujala, P.: Bayesian network modeling of port state control inspection findings and ship accident involvement. Expert Syst. Appl. **41**(4), 1632–1646 (2014)
24. Heij, C., Knapp, S.: Predictive power of inspection outcomes for future shipping accidents–an empirical appraisal with special attention for human factor aspects. Marit. Policy Manag. 1–18 (2018)
25. Grbić, L., Ivanišević, D., Čulin, J.: Detainable Maritime Labour Convention 2006-related deficiencies found by Paris MoU authorities. Pomorstvo **29**(1), 52–57 (2015)
26. Bateman, S.: Maritime security and port state control in the Indian Ocean region. J. Indian Ocean Reg. **8**(2), 188–201 (2012)
27. Cariou, P., Mejia Jr., M.Q., Wolff, F.C.: On the effectiveness of port state control inspections. Transp. Res. Part E Logist. Transp. Rev. **44**(3), 491–503 (2008)
28. Cariou, P., Wolff, F.C.: Do port state control inspections influence flag-and class-hopping phenomena in shipping? J. Transp. Econ. Policy **45**(2), 155–177 (2011)
29. Fan, L., Luo, M., Yin, J.: Flag choice and port state control inspections—empirical evidence using a simultaneous model. Transp. Policy **35**, 350–357 (2014)
30. Titz, M.A.: Port state control versus marine environmental pollution. Marit. Policy Manag. **16**(3), 189–211 (1989)
31. Bang, H.S., Jang, D.J.: Recent developments in regional memorandums of understanding on port state control. Ocean. Dev. Int. Law **43**(2), 170–187 (2012)
32. Liou, S.T., Liu, C.P., Chang, C.C., Yen, D.C.: Restructuring Taiwan's port state control inspection authority. Gov. Inf. Q. **28**(1), 36–46 (2011)
33. Knapp, S., Franses, P.H.: Econometric analysis on the effect of port state control inspections on the probability of casualty: can targeting of substandard ships for inspections be improved? Mar. Policy **31**(4), 550–563 (2007)
34. Knapp, S., Van de Velden, M.: Visualization of differences in treatment of safety inspections across port state control regimes: a case for increased harmonization efforts. Transp. Rev. **29**(4), 499–514 (2009)
35. Knapp, S., Franses, P.H.: Econometric analysis to differentiate effects of various ship safety inspections. Mar. Policy **32**(4), 653–662 (2008)
36. Heij, C., Bijwaard, G.E., Knapp, S.: Ship inspection strategies: effects on maritime safety and environmental protection. Transp. Res. Part D Transp. Environ. **16**(1), 42–48 (2011)
37. Li, K., Tapiero, C.: Strategic naval inspection for port safety and security. Doctoral dissertation, New York University (2010)
38. Qu, X., Meng, Q., Li, S.: Ship collision risk assessment for the Singapore strait. Accid. Anal. Prev. **43**, 2030–2036 (2011)
39. Li, S., Meng, Q., Qu, X.: An overview of maritime waterway quantitative risk assessment models. Risk Anal. **32**(3), 496–512 (2012)

Chapter 25
Dynamic Pricing Model for Container Slot Allocation Considering Port Congestion

Tingsong Wang and Man Li

Abstract This paper studies a dynamic pricing problem for container slot allocation by considering port congestion. To solve the slot allocation with dynamic pricing issue, a one-phase allocation model is proposed to formulate this problem. And, a chance-constrained method is applied to define that the slots reserved for contract shippers can be efficiently used, namely the probability of the slots allocated to contract shippers exceed the actual demand is less than a given parameter.

25.1 Introduction

Recent years, with rapid development of economic globalization, the maritime industry develops steadily, among which the container transportation gets the most rapid growth for its large scale, security, convenience for multimodal transport and so on. In 2015, the total container trade volume amounted in 175 million TEUs [1]. Containerized cargoes of different shippers are transported by shipping lines between their origin ports and destination ports (simply noted as O-D pairs). With the fierce competition in market, it is vital for shipping lines to assign its finite ship capacity resource measured by TEUs to container slot demand in different O-D pairs, to complete customer's demand efficiently as well as maximize the transportation benefits.

Another issue should be concerned by the shipping line is cargo's call for transit time, and we call this category of cargo as time-sensitive cargo. With the improvement of people's living standard, more attention are paid on products' quality, especially for perishable products. In order to keep the quality of the time-sensitive cargoes, customers expect to obtain them as soon as possible, which indicates that time-sensitive cargoes are expected to be expressly delivered; even the payment for such express delivery is high. The high payment motivates a shipping company to receive the

T. Wang (✉) · M. Li
School of Economics and Management, Wuhan University, Wuhan 430072,
People's Republic of China
e-mail: emswangts@whu.edu.cn

M. Li
e-mail: liman@whu.edu.cn

© Springer Nature Singapore Pte Ltd. 2019
X. Qu et al. (eds.), *Smart Transportation Systems 2019*, Smart Innovation,
Systems and Technologies 149, https://doi.org/10.1007/978-981-13-8683-1_25

delivery of time-sensitive cargoes. Moreover, if the cargoes are delivered to shippers before the due date of shipping, the shippers can have more time to deliver the product to customers before expiration with high quality to obtain extra reward; accordingly, the actual freight rate should increase. Of course, if the delivery of cargoes is delayed, shipping company has to be penalized. Hence, whether time-sensitive cargoes can be delivered on time or not is a key issue to be considered by the shipping company. The delivery time consists of shipping time on sailing and staying time in port, which mainly includes the waiting time and service time for loading/unloading at port. The sailing time can be controlled by adjusting the vessel's sailing speed without considering the influence of extreme conditions. However, the staying time in port is not easy to predict by the shipping line due to some uncontrollable factors in terminal operations. The most common factor is port congestion, caused by schedule unreliability in terminals [2], source limitation such as berths, quay cranes, and transport trucks [3, 4], ship collisions or ship groundings [5], and so on, all of which make the ships have to wait on anchor point. Therefore, when shipping line intends to sign a contract of shipping time-sensitive cargoes with the shippers, port congestion is an issue cannot be ignored in the estimation of delivery time because the shipping line needs to leave enough suffer time for the delay at port.

Therefore, the container slot allocation problem of time-sensitive cargoes proposed in this paper is different with that for general cargoes, which indicates that the methodologies in the existing researchers studied for general cargoes cannot be directly applied for the proposed problem in this paper. We need to develop a new one for our problem, which is the work we intend to make an effort to.

The remainder of this paper is organized as follows. Section 25.2 is a review of previous research. Section 25.3 elaborates the container slot allocation of time-sensitive cargo considering time limit and port congestion. Section 25.4 develops the two models for the proposed problem, which are One-phase Allocation Model and Two-phase Allocation Model, respectively. Solution algorithm is proposed in Sect. 25.5, in which chance-constrained programming (CCP) method is introduced to solve the models. Section 25.6 uses a numerical example to evaluate the models and solution algorithm proposed in the study. Finally, Sect. 25.7 concludes the study and provides recommendations for future work.

25.2 Literature Review

Slot allocation issue has been widely studied for decades. Maragos [6] made the first step to analyse the feature of liner shipping and considered the problem of the dynamic slot allocation and pricing in both single-segment and multi-segment container shipping. While a very important problem in shipping operation management, i.e. the empty container repositioning problem caused by region trade imbalance was not considered. Feng and Chang [7], and Song and Carter [8] solved the empty container repositioning problem within seaborne shipping networks, while no laden container routing was done explicitly. Feng and Chang [9] studied the optimal slot allocation problem serving a specific shipping service route for ocean carriers, and took into account the empty containers and laden one's allocation. However, in their papers, the demand has been assumed known and deterministic, while in reality, the demand generally fluctuates and the unknown and uncertain demand is more practical. Considering the demand uncertainty, Bu et al. [10] developed two stochastic programming models on capacity allocation with and without empty containers transportation involved, which were solved by the method of robust optimization. Wang [11] considered the stochastic resource allocation problem for containerized cargo transportation with uncertain capacities and network effects, and provided theoretical results about the proposed constrained stochastic programming model.

Considering the dynamic pricing problem, Feng and Xiao [12] addressed the integrated dynamic pricing and capacity allocation problem for perishable products, in which they assumed that the supplier sells the same products to different micro-markets at distinct prices. Taudes and Rudloff [13] proposed a pricing and inventory control model with a two-period linear demand model, proving that the optimal joint pricing/inventory policy for the replenishment opportunity after the first period is a base-stock list-price policy. Zhu [14] studied a single-item periodic-review model for dynamic pricing problem with returns and expediting. Lee [15] studied a periodic-review pricing and inventory replenishment problem with stochastic demands in multiple periods. Liu et al. [16] proposed a joint slot allocation and dynamic pricing model with demand uncertainties in the container sea–rail multimodal transport system. It can be seen that the dynamic pricing problem are usually a multi-phase issue with distinct prices.

It is noted that the existing literature made the most effort on slot allocation and dynamic pricing of ordinary cargo, while few studies can be found about slot allocation for time-sensitive cargo. According to Panayides and Song [17] and Wang and Meng [18], the 'time factor' is a fundamental requirement of practical liner shipping networks, in which port congestion plays an important role. Meng and Wang [19], Wang and Meng [18] and Meng et al. [20] considered transit time to solve the container paths and network design problem. Wang and Meng [21] proposed a mixed-integer nonlinear stochastic model to hedge against uncertain container handling times and port congestion. Wang et al. [22] took port time windows in their non-linear model to deal with ship route schedule design problem. Readers can refer to Meng et al. [23] for further information on the cargo allocation and scheduling problem. While all of them pay more attention on network design, rather than the container slot allocation problem. As for time-sensitive cargo shipping demand, Wang et al. [24, 25] considered the transit-time-sensitive demand which was assumed to be a decreasing continuous function of transit time, to optimal containership schedule as well as the total profit. Also, path schedule is the main concern in those papers. However, we intend to pay more attention to slot allocation with dynamic pricing problem, which is distinct to the previous studies.

In conclusion, to the best of our knowledge, the container slot allocation problem with dynamic pricing for time-sensitive cargoes has been hardly studied yet. For time-sensitive cargoes, the design of new pricing pattern for freight rate is the main issue we concern, thus we make a correlation between freight rate and delivery time, in which port congestion is taken into account as the major influence factor to delivery time. In addition, most existing researches take the uncertainty of demand of cargo into account and regard the market of cargo as two categories: contract cargo and spot cargo, but rarely take the contract market and spot market of cargoes as a whole system. Therefore, this paper will study the container allocation problem for time-sensitive cargoes in which the demand uncertainty and empty transportation caused by imbalanced region trade are two key issues to be involved in our considerations, and a new pricing pattern is designed considering port congestion.

25.3 Problem Statement

The shipping company operates a shipping line, which calls at total m ports in sequence in an itinerary line. The port rotation in the line can be expressed as: $P_1 \rightarrow P_2 \rightarrow P_3 \rightarrow \cdots \rightarrow P_m \rightarrow P_1$ (Wang et al. [26]). The shipping line allocates the finite capacity denoted by Q to different shippers to satisfy their demands and maximize the total freight revenue of shipping line.

Shipping company divides the market into two groups: contract sale market and free sale market (or spot market). On one hand, in contract market, the larger shippers have a regular, steady, and large demand. To obtain a stable income, shipping line usually reserves most container slots for the contract customers with a lower freight rate for their bargaining power. Meanwhile, the demand on contract market is uncertain, and hence the shipping line needs to predict the demand according to the historical data. On the other hand, in the spot market, the scattered shippers can book slots during t periods of the whole freight solicitation time T, and the greater value of t indicates the closer to the final deadline for book date and the less sensitive of shippers' demand to price changes, so the price will always go up over time. With the growth of the price, free sale demand varies inversely with changes which are assumed to be a simple linear relationship. Moreover, considering the trade imbalance among ports, there may be empty container demand in some ports, resulting in empty container reposition, which produces cost for the shipping company and is an issue we must consider when making slots allocation determination. Therefore, the shipping company allocates the Q slots in three aspects. First, a great proportion of the slots are sold in advance with a lower price as a series contract sale negotiated with large shippers in contract sale market. Second, considering empty container demand caused by trade imbalance among ports, shipping company must set aside some slots to fulfil the empty container demand of some ports. Third, the remaining slots are sold to the scattered shippers who have no bargaining power in the spot market freely with a series relatively high price associated with booking period.

Considering the delivery requirement of time-sensitive cargo on due time, a new price mechanism is designed, in which we make a correlation between freight rate and delivery time, and a penalty/incentive factor is introduced for delivered delay or advance each one day, respectively. Hence a basis price is set up for time-sensitive cargo delivered just on the agreed delivery time $U^{(p^i, p^j)}$ for each loaded container O-

D pair (p^i, p^j), however, when cargo is delivered in advance, the shipping company will charge an extra fee; on the contrary, if cargo delivered exceeds the time limit, the company should pay a certain penalty for the overdue time. Thus, the actual freight rate in final will be the basis price minus a penalty price which equals the penalty factor multiplies the overdue time, or plus an incentive price which equals the incentive factor multiplied the advance delivery time. As to the actual delivery time, it is mainly concerned with sailing time on board, and the waiting and loading/unloading time on loaded port considering port congestion. As mentioned before, the most common uncontrollable factor is port congestion, which further impacts the total delivery time as well as the actual freight rate. Given that the complexity of influence factors on port congestion, this paper considers queuing theory, regarding the port system as an M/M/c system to construct slot allocation models. Suppose that there is a finite number of cargo handling equipment in each port, which have independent service time which follows a negative exponential distribution, and the ships arrival with Poisson process and the time interval between two successive arrivals follows a negative exponential distribution. Although in reality the arrival time of ships are given at least one week in advance, while the information is unknown to the shipping line. Thus, when the shipping line predicts the waiting time at port to determine the negotiated delivery time with shippers, the service system can be regarded as an M/M/c queuing system. According to the queuing theory, ship's dwell time in a port, including waiting to be served and service time, also follows a negative exponential distribution.

Traditional studies on slot allocation problem usually divide the allocation process into two stages, and assign as many as possible container slots to large shippers on contract market to obtain a steady income. However, scattered shippers on spot market are the ones that can provide higher freight rate to get instant service, and certainly high yield means high risk for shipping line. Therefore, it indicates that we need to determine a proper proportion in the allocation scheme, other than giving an absolute preference to contract shippers or determining a proportion artificially. Accordingly, this paper proposes to take the three parts as a whole system, consider the contract container slots, scattered container slots and empty container slots together to make the total expected revenue maximize.

25.4 Model Development

With the consideration of container slot demand uncertainty, and the new price mechanism caused by the attribute of time-sensitive cargo, there is a need to build a mathematical model for the container shipping company to handle the issue so as to maximize the revenue. Through the new price mechanism, the actual price of time-sensitive cargo for contract and scattered shippers can be expressed by Eqs. (25.1) and (25.2), respectively, in which the items in parentheses represent the difference between actual delivery time (the sum of sailing time and service time) and negotiated time for cargo from p^i to p^j.

$$p_a^{(p^i,p^j)} = p_{a^0}^{(p^i,p^j)} - (\Delta t^{(p^i,p^j)} + w^{p^j} - U^{(p^i,p^j)}) \times e, \quad \forall (p^i,p^j) \in M \quad (25.1)$$

$$p_{bt}^{(p^i,p^j)} = p_{b^0t}^{(p^i,p^j)} - (\Delta t^{(p^i,p^j)} + w^{p^j} - U^{(p^i,p^j)}) \times e, \quad \forall (p^i,p^j) \in M, \forall t \quad (25.2)$$

According to the description of the first method, we consider the slot allocation as a whole to construct model I. The model I is a One-phase Allocation Model, in which the slot allocation system is taken as an entirety, and the objective is to maximize the whole revenue. The model is formulated as a chance-constrained stochastic integer programming with dynamic pricing.

$$\max Z = \sum_{(p^i,p^j)\in M} p_a^{(p^i,p^j)} x_a^{(p^i,p^j)} + \sum_{t=1}^{T} \sum_{(p^i,p^j)\in M} p_{bt}^{(p^i,p^j)} x_{bt}^{(p^i,p^j)} - \sum_{(p^i,p^j)\in M} c^{(p^i,p^j)} x_c^{(p^i,p^j)} \quad (25.3)$$

subject to

$$x_{bt}^{(p^i,p^j)} = \alpha_t^{(p^i,p^j)} - \beta_t^{(p^i,p^j)} p_{b^0t}^{(p^i,p^j)}, \quad \forall (p^i,p^j) \in M, \forall t \quad (25.4)$$

$$p_{a^0}^{(p^i,p^j)} \leq p_{b^0t}^{(p^i,p^j)} \leq P_U^{(p^i,p^j)}, \quad \forall (p^i,p^j) \in M, \forall t \quad (25.5)$$

$$\eta_{al} + \sum_{t=1}^{T} \eta_{btl} + \eta_{cl} \leq Q, \quad \forall l = 1, 2, \ldots, m \quad (25.6)$$

$$\Pr(x_a^{(p^i,p^j)} \leq D_a^{(p^i,p^j)}) \geq 1 - \alpha, \quad \forall (p^i,p^j) \in M \quad (25.7)$$

$$\sum_{(p^i,p^j)\in M} x_c^{(p^i,p^j)} \geq E^{p^j}, \quad \forall p^j \in \Omega \quad (25.8)$$

$$x_c^{(p^i,p^j)} \leq ES^{p^i}, \quad \forall p^i \in \Omega \quad (25.9)$$

$$x_c^{(p^j,p^k)} = 0, \text{ when } E^{p^j} > 0, \quad \forall p^j, p^k \in \Omega \quad (25.10)$$

$$x_a^{(p^i,p^j)}, x_{bt}^{(p^i,p^j)}, x_c^{(p^i,p^j)} \in N \cup \{0\}, \quad \forall (p^i,p^j) \in M \quad (25.11)$$

Equation (25.3) is the objective function of this model, in which the first term is the profit of shipping contract cargoes, and the second term is the revenue of cargos from scattered shippers in all booking periods of free sale, finally, the last term is the cost of empty containers transportation, which must be deducted from the company's total revenue. The set of constraints (25.4) indicates the linear relation between price and demand in free market. The set of constraints (25.5) ensures that the basis price in free market is in any period cannot be less than the basis price of contract sale, and cannot be more than a price upper limit on each O-D pair. The set of constraints (25.6) requires that the total number of slot allocated to the contract shippers, the scattered shippers and empty containers cannot exceed the capacity of the ship. The set of constraints (25.7) is a chance constraint, which defines that the slots reserved for contract shippers can be efficiently used, namely the probability of the slots allocated to contract shippers exceed the actual demand is less than α. Constraint set (25.8) presents the empty container slot allocation from each p^i to p^j cannot be less than the empty container demand of p^j. Constraint set (25.9) presents the empty containers transporting from p^i to p^j cannot be more than the available empty containers in port p^i. While constraint set (25.10) indicates that the empty container cannot be transported from p^j to the other ports when p^j has an empty container demand. Constraint (25.11) is an integer constraint of the decision variables.

References

1. UNCTAD: Review of maritime transportation 2015. In: Paper presented at the united nations conference on trade and development, New York, Geneva. http://unctad.org/en/PublicationsLibrary/rmt2016_en.pdf, Accessed 15 Dec 2016
2. Brooks, M.R., Schellinck, T.: Measuring port effectiveness: what really determines cargo interests' evaluations of port service delivery? Marit. Policy Manag. 42(7), 699–711 (2015)
3. Wan, Y., Yuen, C. L., Zhang, A.: Effects of hinterland accessibility on us container port efficiency. Int. J. Shipp. Transp. Logist. 6(4) (2014)
4. Shang, X.T., Cao, J.X., Ren, J.: A robust optimization approach to the integrated berth allocation and quay crane assignment problem. Trans. Res. Part E: Logist. Transp. Rev. 94, 44–65 (2016)
5. Jiang, C., Wan, Y., Zhang, A.: Internalization of port congestion: strategic effect behind shipping line delays and implications for terminal charges and investment. Marit. Policy Manag. 1–19 (2016)
6. Maragos, S.A.: Yield Management for the Maritime Industry. Massachusetts Institute of Technology (1994)
7. Feng, C., Chang, C.: Empty container reposition planning for intra-Asia liner shipping. Marit. Policy Manag. 35(5), 469–489 (2008)
8. Song, D., Carter, J.: Empty container repositioning in liner shipping. Marit. Policy Manag. 36(4), 291–307 (2009)
9. Feng, C.M., Chang, C.H.: Optimal slot allocation with empty container reposition problem for Asia ocean carriers. Int. J. Shipp. Transp. Logist. 2(1), 22–43 (2010)
10. Bu, X.Z., Zhao, Q.W., Huang, Q., et al.: Optimal capacity allocation model of ocean shipping container revenue management considering empty container transportation. Chin. J. Manag. Sci. 13(1), 71–75 (2005)
11. Wang, X.: Stochastic resource allocation for containerized cargo transportation networks when capacities are uncertain. Transp. Res. Part E: Logist. Transp. Rev. 93, 334–357 (2016)
12. Feng, Y.Y., Xiao, B.C.: Integration of pricing and capacity allocation for perishable products. Eur. J. Oper. Res. 168(1), 17–34 (2006)
13. Taudes, A., Rudloff, C.: Integrating inventory control and a price change in the presence of reference price effects: a two-period model. Math. Methods Oper. Res. 75(1), 29–65 (2012)
14. Zhu, S.X.: Joint pricing and inventory replenishment decisions with returns and expediting. Eur. J. Oper. Res. 216(1), 105–112 (2012)
15. Lee, J.: Dynamic pricing inventory control under fixed cost and lost sales. Appl. Math. Model. 38(2), 712–721 (2014)

16. Liu, D., Yang, H.: Joint slot allocation and dynamic pricing of container sea–rail multimodal transportation. J. Traffic Transp. Eng. **2**(3), 198–208 (2015)
17. Panayides, P., Song, D.: Maritime logistics as an emerging discipline. Marit. Policy Manag. **40**(3), 295–308 (2013)
18. Wang, S., Meng, Q.: Liner shipping network design with deadlines. Comput. Oper. Res. **41**, 140–149 (2014)
19. Meng, Q., Wang, S.: Optimal operating strategy for a long-haul liner service route. Eur. J. Oper. Res. **215**(1), 105–114 (2011)
20. Meng, Q., Wang, T., Wang, S.: Multi-period liner ship fleet planning with dependent uncertain container shipment demand. Marit. Policy Manag. **42**(1), 43–67 (2015)
21. Wang, S., Meng, Q.: Robust schedule design for liner shipping services. Transp. Res. Part E: Logist. Transp. Rev. **48**(6), 1093–1106 (2012)
22. Wang, S., Alharbi, A., Davy, P.: Liner ship route schedule design with port time windows. Transp. Res. Part C: Emerg. Technol. **41**, 1–17 (2014)
23. Meng, Q., Wang, S., Andersson, H., Thun, K.: Containership routing and scheduling in liner shipping: overview and future research directions. Transp. Sci. (2013). https://doi.org/10.1287/trsc.2013.0461
24. Wang, S., Meng, Q., Liu, Z.: Containership scheduling with transit-time-sensitive container shipment demand. Transp. Res. Part B: Methodol. **54**(3), 68–83 (2013)
25. Wang, S., Meng, Q., Lee, C.Y.: Liner container assignment model with transit-time-sensitive container shipment demand and its applications. Transp. Res. Part B: Methodol. **90**, 135–155 (2016)
26. Wang, T., Meng, Q., Wang, S., et al.: Risk management in liner ship fleet deployment: a joint chance constrained programming model. Transp. Res. Part E **60**(4), 1–12 (2013)

Chapter 26
Parcel Sorting Optimization in Double-Layer Automatic Sorting System

Zheyi Tan, Lu Zhen, Liyang Xiao and Qian Sun

Abstract Fully automated sorting equipment plays a crucial role in modern distribution networks of the parcel service industry, and the efficiency of the sorting scheme is crucial for the sorting equipment. This study optimizes the sorting scheme for a type of double-layer sorting equipment, which adapts "component-sorting" strategy. "Component-sorting" refers to the method of sorting multiple parcels sent to the same destination by the same tray. The problem is formulated as an integer programming model. Then, we propose an efficient variable neighbor taboo search to solve the problem within a short time. The experimental results show that the proposed solution method is suitable for solving the problem of interest.

26.1 Introduction

With the rapid development of e-commerce, the burden of distribution centers has become heavier than ever [1]. To improve the efficiency of sorting, much more companies are adopting automated sorting equipment [2, 3]. In such a conveyor system, workers remove the parcels from the inbound trailer and place them on separate trays of the conveyor until they are transported to outbound trucks dedicated to the parcel's destination [4, 5]. The conveyor is the heart of the distribution center and there is no doubt that improving the efficiency of the conveyor is critical to the overall operation of distribution center.

The literature of automatic sorting systems mainly focused on two aspects: hardware and software. Studies of hardware aimed at improving the sorting equipment's structure, such as the optimal layout of the double-layer sorting equipment [6], merging conveyors in sequential zone picking systems [7], parking the racks of KIVA [8].

As for the literature stream of software, many scholars optimized the sorting scheduling scheme. Hou et al. [9] analyzed the problem of order allocation in detail,

Z. Tan (✉) · L. Zhen · L. Xiao · Q. Sun
School of Management, Shanghai University, Shanghai, China
e-mail: 395549601@qq.com

L. Zhen
e-mail: lzhen@shu.edu.cn

© Springer Nature Singapore Pte Ltd. 2019
X. Qu et al. (eds.), *Smart Transportation Systems 2019*, Smart Innovation,
Systems and Technologies 149, https://doi.org/10.1007/978-981-13-8683-1_26

established a mathematical model with the goal of equalization of sorting task assignment, and proved its effectiveness by examples, and got the most excellent job allocation plan. Zhang and Wu [10] studied the parallel automatic sorting system and discussed the problem of confluence of goods on the conveyor, and reduced sorting time by shifting the time when the item starts sorting.

This study evaluated the sorting and dispatching scheme of the double-layered cross-belt sorting equipment in the parcel distribution center. Compared with the single-layer sorting equipment, the double-layer sorting equipment has higher utilization rate of the sorting compartment, and the parcels can be sorted by suitable sorting compartments. This study optimizes the sorting scheme for double-layer sorting equipment, which adapts "component-sorting" strategy. "Component-sorting" refers to the way that parcels for the same destination are allowed to be sorted by one tray. The use of "component sorting" can significantly improve the utilization rate of the tray and the sorting distance can be shortened, thereby improving the sorting efficiency.

In order to further improve the sorting efficiency, based on the research of double-layer sorting equipment, this paper puts forward the problem of destination assignment of double-layer sorting equipment considering "component sorting", establishes a mathematical model, and designs a variable neighborhood taboo search to solve it. The experimental results verify the effectiveness and superiority of the algorithm.

26.2 Problem Background

The distribution center mainly uses double-layered cross-belt sorting equipment. The upper conveyor of the sorting equipment runs in a counterclockwise direction, and the lower conveyor runs in a clockwise direction. After the parcel enters the distribution center, it is sent to the packing machine by the inbound trailer. First, it is pre-sorted according to the parcel number, and then the parcel is sent to the conveyor according to the sorting profile. Each parcel is sorted by a cross-belt tray. When the parcel is delivered to the designated sorting compartment, the cross-belt tray unloads the parcel, and the parcel enters the designated material bin from the spiral chute, ending the sorting of the parcel. When the parcel in the material box reaches a certain amount, the outbound trailer will load the parcel into the outbound truck for distribution. Each sorting device has completed the sorting of the batch of parcels before the next batch of parcels arrives.

26.2.1 Sorting-Profile Selection Problem

In order to simplify the problem and combine it with the actual situation, this paper introduces the concept of sorting profile, that is, the number of parcels of each

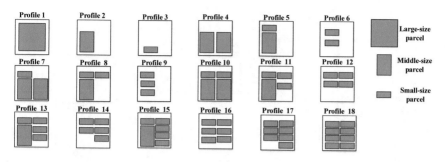

Fig. 26.1 An example of sorting profile for a parcel

specification loaded on each tray must be set with a predetermined value. Since in the sorting process, the parcel is required to place the order face up. It is convenient to scan the parcel information, and the QR code of the parcel is generally attached to the largest area, so each parcel cannot be stacked. In each sorting profile, the number of parcels that can be placed is fixed, and in real life, the parcels are usually classified according to specifications: oversized, large size, medium size, and small size, wherein the oversized parcels are separately classified. Therefore, there are only three types of parcels sorted in this article. The size of the large parcel is 0.7×0.7 m. The size of the middle parcel and the small parcel are, respectively, 0.6×0.4 m and 0.4×0.2 m. The maximum size of the transport parcel for the tray is 0.8×0.8 m, we can know that all the sorting profiles of one tray are as shown in Fig. 26.1.

26.2.2 Destination Assignment Problem with Multiple Parcel Conveyors

The top view of the sorting device is shown in Fig. 26.2. The cross-belt sorting equipment has a fixed tray area, and the area of the trays occupied by each parcel is mostly different. By considering "component sorting", the tray utilization rate can be improved, and the sorting effectiveness can be improved. If "component sorting" is carried out, the following conditions must be met: (1) the destination of parcel is the same; (2) the parcel is sorted by the same conveyor belt.

For the characteristics of the double-layer sorting equipment studied in this paper, the objective function of this problem cannot only minimize the total transport distance. If this is the case, the workload of the two conveyors cannot be balanced well. The objective function of the model in this paper is to minimize the larger total sorting distance between two conveyors. Therefore, according to the definition of "component sorting" and the feature of double-layer sorting equipment, this paper studies the destination assignment problem of double-layer sorting equipment considering "component sorting", establishes a 0–1 planning model.

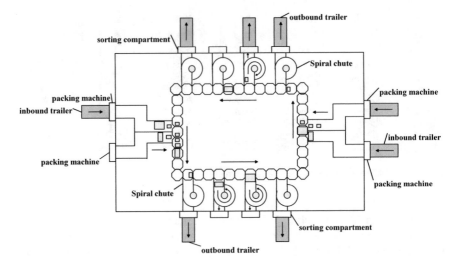

Fig. 26.2 Top viewer of the double-layered cross-belt sorting equipment

26.2.3 Notations

Before presenting the model, notations on parameters and variables used in the model are listed as follows. For the convenience of understanding the notations, we use Latin letters and Greek letters to denote the parameters (indices, sets) and the variables, respectively.

Indices and Sets

D set of destinations, indexed by d.
P set of sorting profiles, indexed by p.
I set of parcels, indexed by i.
S set of parcels' sizes, indexed by s.
E set of sorting compartments, indexed by e.
K set of conveyors, indexed by k.
J set of trays, indexed by j.
I_d set of parcels for destination d.

Parameters

$t_{e,k}$ the sorting distance of the parcel from the conveyor k sorting by the sorting compartment e. $h_{i,d}$ equals one if the destination of parcel i is d, and zero otherwise. $f_{i,s}$ equals one if the size of parcel i is s, and zero otherwise. n_s the number of parcels' size is s. $g_{p,s}$ in the sorting profile p, the number of parcels' size is s.

Variables

$\alpha_{j,p}^d$ binary, equals one if tray j uses the sorting profile p to sort the parcels with destination d, and zero otherwise.

$\beta_{j,k,e}^d$ binary, equals one if tray j is sorted from sorting compartment e and transported by conveyor k, and zero otherwise.

$\gamma_{i,j}^d$ binary, equals one if parcel i for destination d is sorted by tray j, and zero otherwise.

$\delta_{d,e}$ binary, equals one if parcels for destination d is sorted from the sorting compartment e, and zero otherwise.

$\theta_{j,d}$ binary, equals one if parcels for destination d are sorted by tray j, and zero otherwise.

n the largest total sorting distance of two conveyors.

26.2.4 Mathematical Model

$$min\ n \tag{26.1}$$

$$s.t. \sum_{d \in D} \sum_{p \in P \cup \{0\}} \alpha_{j,p}^d = 1 \ \forall j \in J \tag{26.2}$$

$$\sum_{d \in D} \sum_{e \in E \cup \{0\}} \sum_{k \in K} \beta_{j,k,e}^d = 1 \ \forall j \in J \tag{26.3}$$

$$\sum_{j \in J} \gamma_{i,j}^d = 1 \ \forall d \in D, \forall i \in I_d \tag{26.4}$$

$$\sum_{d \in D} \delta_{d,e} \leq 1 \ \forall e \in E \tag{26.5}$$

$$\sum_{e \in E} \delta_{d,e} \geq 1 \ \forall d \in D \tag{26.6}$$

$$\theta_{j,d} = \sum_{e \in E} \sum_{k \in K} \beta_{j,k,e}^d \ \forall d \in D, \forall j \in J \tag{26.7}$$

$$\gamma_{i,j}^d \leq \theta_{j,d} \ \forall i \in I_d, \forall j \in J, \forall d \in D \tag{26.8}$$

$$\alpha_{j,p}^d \leq \theta_{j,d} \ \forall d \in D, \forall j \in J, \forall p \in P \tag{26.9}$$

$$\sum_{j \in J} \sum_{k \in K} \beta_{j,k,e}^d \leq |J| \delta_{d,e} \ \forall d \in D, \forall e \in E \tag{26.10}$$

$$\sum_{i \in I_d} f_{i,s} \gamma_{i,j}^d \leq g_{p,s} \alpha_{j,p}^d + |I| \left(1 - \alpha_{j,p}^d\right) \ \forall d \in D, \forall j \in J, \forall s \in S, \forall p \in P \cup \{0\} \tag{26.11}$$

$$\sum_{i \in I_d} f_{i,s} \gamma_{i,j}^d + |I| \left(1 - \alpha_{j,p}^d\right) \geq g_{p,s} \alpha_{j,p}^d \ \forall d \in D, \forall j \in J, \forall s \in S, \forall p \in P \cup \{0\} \tag{26.12}$$

$$1 - \sum_{k \in K} \beta_{j,k,0}^d \geq \gamma_{i,j}^d \ \forall d \in D, \forall i \in I_d, \forall j \in J \tag{26.13}$$

$$\sum_{k \in K} \beta_{j,k,0}^d = \alpha_{j,0}^d \ \forall d \in D, \forall j \in J \tag{26.14}$$

$$\sum_{d \in D} \sum_{j \in J} \sum_{p \in P \cup \{0\}} g_{p,s} \alpha_{j,p}^d = n_s \ \forall s \in S \tag{26.15}$$

$$n \geq \sum_{d \in D} \sum_{e \in E} \sum_{j \in J} t_{e,k} \beta_{j,k,e}^d \ \forall k \in K \tag{26.16}$$

$$\alpha_{j,p}^d \in \{0, 1\} \ \forall d \in D, \forall j \in J, p \in P \tag{26.17}$$

$$\beta_{j,k,e}^d \in \{0, 1\} \ \forall j \in J, \forall e \in E, \forall k \in K, \forall d \in D \tag{26.18}$$

$$\gamma_{i,j}^d \in \{0, 1\} \ \forall d \in D, \forall i \in I, \forall j \in J \tag{26.19}$$

$$\delta_{d,e} \in \{0, 1\} \ \forall d \in D, \forall e \in E \tag{26.20}$$

$$\theta_{j,d} \in \{0, 1\} \ \forall j \in J, \forall d \in D \tag{26.21}$$

Objective (26.1) minimizes the largest total sorting distance of two conveyors. Constraint (26.2) ensures that only one sorting profile can be selected for each tray. Constraint (26.3) makes sure that a tray can be either enabled or disabled. If enabled, it can only be transported on one conveyor, and only one sorting compartment can be selected for its contained parcels. Constraint (26.4) guarantees that each parcel is sorted by exactly one tray. Constraint (26.5) ensures that a sorting compartment is allocated to at most one destination. Constraint (26.6) ensures that the parcels of each destination are allocated to at least one sorting compartment. Constraint (26.7) links variables $\theta_{j,d}$ and $\beta_{j,k,e}^d$. Constraint (26.8) guarantees the consistency between parcel's destination and the tray destination. Constraint (26.9) indicates the relationship between variables $\alpha_{j,p}^d$ and $\theta_{j,d}$. Constraint (26.10) ensures that if sorting compartment e is not allocated to sort destination d's parcels, then no tray transports destination d's parcels to sorting compartment e. Constraints (26.11) and (26.12) connect variables $\gamma_{i,j}^d$ and $\alpha_{j,p}^d$. Constraints (26.13) and (26.14) indicate that only enabled trays can sort parcels. Constraint (26.15) means that for each size, all parcels are sorted by trays. Constraint (26.16) determines the largest total sorting distance of two conveyors. Finally, constraints (26.17)–(26.21) define decision variables.

26.3 Solution Procedures

Based on the problem of destination assignment of double-layer sorting equipment, this study considers a double-layer sorting equipment destination assignment problem with "component-sorting" strategy. Since the problem of destination assignment of double-layer sorting equipment is NP-hard problem [6], the problem raised in this paper is also NP-hard problem. For the large-size instances, it is difficult to obtain the optimal solution in a short time by using the exact solution algorithm, so this paper solves the problem by using the double variable neighborhood taboo search.

Double variable neighborhood taboo search (VNTS^2) is a meta-heuristic method combining the variable neighborhood search (VNS) and taboo search (TS). The neighborhood structure of algorithm does not search according to the fixed pattern, which can jump out the local optimum more effectively and improve the quality of the solution. The main idea of the algorithm is as follows: obtain an initial solution, optimize it by changing the neighborhood search strategy, solve the local optimal solution, and then store the local optimal solution searched to avoid searching for the local maximum again in further search.

(1) Determination of the initial solution

The pros and cons of the initial solution have a great influence on the search performance of the double variable neighborhood taboo search. In order to improve working efficiency and speed of solving, the initial solution is generated as follows: the sorting compartment is allocated according to the number of parcels per destination. Then, the sorting profile for each tray is decided to reduce the number of trays enabled, and we assign to $\alpha_{j,p}^d$, $\gamma_{i,j}^d$ and $\theta_{j,d}$. Then, the destination with more tray enabled is matched to the sorting compartment that is closer to the charter, and then we assign to $\delta_{d,e}$. After determining the sorting compartment, $\beta_{j,k,e}^d$ is assigned.

(2) Variable neighborhood search strategy

Mutation. The point mutation operation is mainly performed for $\delta_{d,e}$, $\alpha_{j,p}^d$, $\gamma_{i,j}^d$, $\beta_{j,k,e}^d$ to perform the mutation operation. If the original value of the decision variable is 0, it turns to 1, and if it is 1, it turns to 0. The remaining variables are then adjusted according to the mathematical model to obtain the solution in this case.

Exchange. The point exchange operation is mainly performed for $\delta_{d,e}$, $\alpha_{j,p}^d$, $\beta_{j,k,e}^d$, for example, if $j_1 \neq j_2$, $k_1 \neq k_2$, $\beta_{j_1,k_1,e}^d = 1$, $\beta_{j_2,k_2,e}^d = 1$, then exchange the conveyor where these two trays are transported by, and let $\beta_{j_1,k_2,e}^d = 1$, $\beta_{j_2,k_1,e}^d = 1$. Finally, the remaining variables are then adjusted according to the mathematical model, and the solution in this case is obtained.

26.4 Computational Experiment

In this part, several groups of artificial instances are randomly generated. All experiments were performed on a computer with 2.4 GHz Intel Core i5 CPU and 256 GB RAM. The algorithm was coded using Visual Studio 2012 and CPLEX 12.6.1. The time limit for all test instances was two hours (7,200 s).

In the following table, I indicates the number of parcels, D indicates the number of destinations. Moreover, *Gap* refers to the difference between the lower bound of the model and the best solution value obtained by CPLEX. $\Delta_{VNTS^2} = \frac{F_{VNTS^2} - F_{CPLEX}}{F_{CPLEX}}$ compares the solutions obtained by VNTS^2 and CPLEX. Table 26.1 shows the performance of VNTS^2 for small-size and large-size instances. From this value, we can

Table 26.1 Performance of VNTS^2 for small-size and large-size instances

ID	I	D	CPLEX			VNTS^2		Δ_{VNTS^2} (%)
			F_{CPLEX}	Gap (%)	t_{CPLEX} (s)	F_{VNTS^2}	t_{VNTS^2} (s)	
1-1	10	4	2	0.00	0.94	2	0.12	0.00
1-2	10	4	4	0.00	0.93	4	0.35	0.00
1-3	10	4	4	0.00	0.95	4	0.27	0.00
2-1	20	6	8	0.00	16.43	8	2.41	0.00
2-2	20	6	7	0.00	15.94	7	1.96	0.00
2-3	20	6	10	0.00	25.75	10	2.52	0.00
3-1	30	8	16	12.50	7200.00	15	5.21	−6.25
3-2	30	8	17	16.67	7200.00	17	4.85	0.00
3-3	30	8	18	16.67	7200.00	17	5.38	−5.56
4-1	40	10	21	33.33	7200.00	21	40.52	0.00
4-2	40	10	20	40.00	7200.00	19	36.93	−5.00
4-3	40	10	21	38.10	7200.00	18	42.89	−14.29

know the difference between the objective function values of VNTS^2 and CPLEX. For small-size instances, as everyone can see it, the speed of VNTS^2 is much faster than CPLEX. In addition, when the number of parcels reaches 30 pieces, the solution time of CPLEX is increasing rapidly. For large-size instances, when the number of parcels is more than 30 pieces, CPLEX cannot find optimal solutions within two hours and the VNTS^2 outperforms CPLEX in both terms of solution quality and efficiency. Therefore, it can be concluded that the VNTS^2 proposed in this study is more suitable for solving this problem.

26.5 Conclusion

In this study, we propose the destination assignment problem of a double-layer sorting equipment considering "component sorting". For this problem, this study designs a kind of algorithm combined taboo search with variable neighborhood search, which can generate a better sorting scheduling scheme in a shorter time to assist the equipment operator in making decisions. Compared with CPLEX, the algorithm has a shorter time to get a solution and can solve large-size instances efficiently. Therefore, VNTS^2 is more suitable for solving this problem.

References

1. Boysen, N., Fedtke, S., Weidinger, F.: Optimizing automated sorting in warehouses: the minimum order spread sequencing problem. Eur. J. Oper. Res. **270**(1), 386–400 (2016)
2. Boysen, N., Fliedner, M.: Scheduling aircraft landings to balance workload of ground staff. Comput. Ind. Eng. **60**(2), 206–217 (2011)
3. Bozer, Y.A., Hsieh, Y.J.: Expected waiting times at loading stations in discrete-space closed-loop conveyors. Eur. J. Oper. Res. **155**(2), 516–532 (2004)
4. Bozer, Y.A., Hsieh, Y.J.: Throughput performance analysis and machine layout for discrete-space closed-loop conveyors. IIE Trans. **37**(1), 77–89 (2005)
5. Khachatryan, M.: Small Parts High Volume Order Picking Systems. Georgia Institute of Technology (2006)
6. Fedtke, S., Boysen, N.: Layout planning of sortation conveyors in parcel distribution centers. Transp. Sci. **51**(1), 3–18 (2017)
7. van der Gaast, J.P., de Koster, M.B.M., Adan, I.J.B.F.: Conveyor merges in zone picking systems: a tractable and accurate approximate model. Transp. Sci. **52**(6), 14–28 (2018)
8. Weidinger, F., Boysen, N., Briskorn, D.: Storage assignment with rack-moving mobile robots in KIVA warehouses. Trans. Sci. **52**(6), 1479–1495 (2018)
9. Hou, J.L., Wu, N., Wu, Y.J.: A job assignment model for conveyor-aided picking system. Comput. Ind. Eng. **56**(4), 1254–1264 (2009)
10. Wu, Y.H., Zhang, Y.G., Wu, Y.Y.: Compressible virtual window algorithm in picking process control of automated sorting system. Chin. J. Mech. Eng. **21**(3), 41–45 (2008)

Chapter 27
Data Completion of Ride-Hailing Service Based on Tensor Factorization

Yan Xia, Ruo Jia, Zhekang Li, Jiayan Zhu, Chenxi Hu, Zhiyuan Liu and Zewen Wang

Abstract This paper adopts the modified CPWOPT (CANDECOMP/PARAFAC Weighted Optimization) to recover the missing traffic speed data collected in the ride-hailing service. The data completion method based on the tensor decomposition is modified by adding factor tensor in the regular terms, which contains the characteristics of weekdays, time periods, and road segments. After evaluating the performance of the method in the ride-hailing data, the results indicate that the method not only increases the accuracy of data completion compared to CPWOPT, but also fills the missing data more reasonably catering to the temporal distribution of traffic speed data.

27.1 Introduction

In the era of big data, the collected vast data is of considerable significance for the analysis of transport systems and infrastructure [1]. In the transportation system, many data collected from the road often have incomplete data due to detector failure or communication interruption. Therefore, recovering missing traffic data has very strong practical significance. Due to the spatial and temporal correlation in the traffic data, there are certain similarities between different road segments, and there are certain similarities between different time periods [1].

Therefore, it is necessary to use the data completion method to mine the spatial and temporal characteristics of the road network data based on the existing sparse data and complete it [2]. Tensor factorization enables us to take the structure into account by effectively capturing the multilinear interactions among multiple latent factors [3]. The method has been successfully applied to various application fields, such as

Y. Xia · R. Jia · Z. Li · Z. Liu (✉) · Z. Wang
School of Transportation, Southeast University, No. 2, Sipailou, Nanjing, People's Republic of China
e-mail: leakeliu@163.com

J. Zhu · C. Hu
Beijing Didi Infinite Technology Development Co., Ltd., Haidian District, Beijing, People's Republic of China

© Springer Nature Singapore Pte Ltd. 2019
X. Qu et al. (eds.), *Smart Transportation Systems 2019*, Smart Innovation, Systems and Technologies 149, https://doi.org/10.1007/978-981-13-8683-1_27

face recognition, image compression [4]. The two most popular tensor factorization frameworks are Tucker [5] and CANDECOMP/PARAFAC (CP) [6].

In the previous studies, the data completion mainly includes two method, the first direction is unfolding the multidimensional tensor into two dimensional and convert the tensor data into a matrix. Liu et al. [7] propose the method of HALRTC (High accurate Low Rank Tensor Completion), which used the trace norm of tensor to construct the tensor completion into a convex optimization problem. Ran et al. [8] employed the HALRTC method to recover missing entries from given entries, considering severe fluctuation of traffic speed data compared with traffic volume. The second direction is directly decomposing the multidimensional tensor. In this area, Acar [4] formulated a CP model called CPWOPT (CP Weighted Optimization) which is developed using a first-order optimization approach to solve the weighted least squares problem. Wang et al. [9] employed Tucker decomposition and low rank latent factor matrices for time slots, grids, roads, and geographical features. To achieve a high accuracy of decomposition. Some researchers focused on combining the Bayesian model with tensor decomposition. Tang et al. [10] adopted the fully Bayesian treatment and Gibbs sampling to achieve automatic hyperparameter tuning and model complexity controlling. Zhao et al. [11] formulated CP factorization using a probabilistic model and a fully Bayesian treatment by incorporating automatic rank determination.

To sum up, the contribution of this paper is as follows: (1) adopt the data completion method based on the tensor decomposition to the ride-hailing GPS trajectories. (2) propose a modified CP decomposition method of data completion.

27.2 Notations and Tensor Factorization

In the calculation of CP decomposition, some widely adopted methods are used [1], which are introduced as follows for the sake of presentation.

27.2.1 Methods of Product

There are several methods of product commonly used in tensor calculation [12], including Kronecker product, Khatri-Rao product, and outer product.

Kronecker product: given a matrix A of size $m_1 \times m_2$ and a matrix B of size $n_1 \times n_2$, then the Kronecker product denoted as "$\otimes$" of matrix A and matrix B is:

$$A \otimes B = \begin{bmatrix} a_{11}B & a_{12}B & \cdots & a_{1m_2}B \\ a_{21}B & a_{22}B & \cdots & a_{2m_2}B \\ \vdots & \vdots & \ddots & \vdots \\ a_{m_11}B & a_{m_12}B & \cdots & a_{m_1m_2}B \end{bmatrix} \tag{27.1}$$

Khatri-Rao product: given a matrix $A = \left(\overrightarrow{a_1}, \overrightarrow{a_1}, \ldots, \overrightarrow{a_k}\right)$ of size $m \times k$ and a matrix $B = \left(\overrightarrow{b_1}, \overrightarrow{b_2}, \ldots, \overrightarrow{b_k}\right)$ of size $n \times k$, then the Khatri-Rao product [13] denoted by "$\odot$" of matrix A and matrix B is

$$A \odot B = \left(\overrightarrow{a_1} \otimes \overrightarrow{b_1}, \overrightarrow{a_2} \otimes \overrightarrow{b_2}, \ldots, \overrightarrow{a_k} \otimes \overrightarrow{b_k}\right) \tag{27.2}$$

Outer product: given a vector $\vec{a} = (1, 2)^T$, vector $\vec{b} = (3, 4)^T$, then the outer product denoted by the symbol "$\circ$" of $\vec{a}$ and $\vec{b}$ is as following:

$$\vec{a} \circ \vec{b} = \vec{a}\vec{b}^T = \begin{bmatrix} 3 & 4 \\ 6 & 8 \end{bmatrix} \tag{27.3}$$

27.2.2 CP Decomposition

The CP decomposition (CANDECOMP/PARAFA decomposition), which is actually a combination of CANDECOMP method and PARAFA method. For example, a third-order tensor can be decomposed into three one-dimensional factor matrices so that the entire tensor data can be replaced by a smaller amount of data. In general, given a tensor of size $n_1 \times n_2 \times n_3$, its results of CP decomposition can be written as follows:

$$\chi \approx \sum_{r=1}^{R} A(:, r) \otimes B(:, r) \otimes C(:, r) \tag{27.4}$$

where the sizes of the matrices A, B, and C are, respectively, $n_1 \times R$, $n_2 \times R$, $n_3 \times R$ which are regarded as factor matrices. The symbol "$\otimes$" represents a tensor product. The estimated values of the tensor χ at the position (i, j, k) are

$$x_{ijk} \approx \sum_{r=1}^{R} a_{ir}b_{jr}c_{kr} \tag{27.5}$$

In fact, Tucker decomposition is another mainstream tensor decomposition method. There are two reasons accounting for using CP decomposition instead of Tucker. First, the structure of CP decomposition is more concise and easier to understand than Tucker's structure. Second, the disadvantages of CP decomposition can be released in some degree in the traffic network data. In the sparse data of traffic, the rank is reasonable to be as large as possible.

27.3 The CPWOPT Method for Sparse Tensor

After the tensor factorization, the recovery matrix closest to the true value is essential to be computed [14]. This section mainly introduces the optimization method called CPWOPT (CP Weighted Optimization) for sparse tensor, which is a weighted optimization method based on tensor decomposition.

27.3.1 Weighting of Tensor

In order to deal with the sparse tensor, it is necessary to obtain the weighting matrix according to the data position of the tensor. Suppose x is the original tensor and z is the tensor recovered after the CP decomposition shown in the following:

$$z_{ijk} = \sum_{r=1}^{R} a_{ir} b_{jr} c_{kr} \tag{27.6}$$

In order to make the recovered tensor approaching the true value, the objective function is computed as:

$$f(ABC) = \sum_{i=1}^{I} \sum_{j=1}^{J} \sum_{k=1}^{K} \left(x_{ijk} - \sum_{r=1}^{R} a_{ir} b_{jr} c_{kr} \right)^2 \tag{27.7}$$

The normal dense matrix can gradually update the parameters a_{ir}, b_{jr}, c_{kr}, etc. by using gradient descent, and terminate the iteration when the objective function is less than a certain value, thereby outputting all the results. However, sparse matrix should be processed specially. Assuming w is the weighting matrix, the size of which is the same as the tensor x, the weighted tensor y is shown as

$$y = w * x$$

$$\tag{27.8}$$

The weighting matrix w is

$$w_{ijk} = \begin{cases} 0 & \text{Data is missing} \\ 1 & \text{Data is not missing} \end{cases} \tag{27.9}$$

The corresponding weighted objective function is

$$f(ABC,) = \sum_{i=1}^{I} \sum_{j=1}^{J} \sum_{k=1}^{K} \left\{ w_{ijk} \left(x_{ijk} - \sum_{r=1}^{R} a_{ir} b_{jr} c_{kr} \right) \right\}^2 \tag{27.10}$$

The objective function is used for optimization. The estimation error of the observed data in the original tensor is computed to update, other unobserved data will not be updated separately. However, the characteristic of the observed value is used for estimating the missing data.

27.3.2 CPWOPT

The data completion method used in this paper is a modified method based on the CPWOPT method. First, the optimization method of CPWOPT is introduced. The basic calculation steps of CPWOPT are as follows:

Step1: Precompute weighted matrix:
$$\mathscr{y} = \mathscr{w} * \mathscr{x}$$

Step2: Conduct CP decomposition for matrix $\mathscr{x}$: $\mathscr{x} \approx [\![A^{(1)}, \dots, A^{(N)}]\!]$, Weighted recovery tensor: $\mathscr{z} = \mathscr{w} * [\![A^{(1)}, \dots, A^{(N)}]\!]$

Step3: Compute objective function:

$$f_w = \| \mathscr{y} - \mathscr{z} \|^2 \tag{27.11}$$

The objective function can be expanded as follows:

$$f_w = \| \mathscr{y} \|^2 - 2 \langle \mathscr{y}, \ \mathscr{z} \rangle + \| \mathscr{z} \|^2 \tag{27.12}$$

Step4: Recalculate the gradient for each decomposed tensor $A^{(n)}$ during each iteration as follows:

$$G^{(n)} = -2Y_{(n)} A^{(-n)} + 2Z_{(n)} A^{(-n)} \tag{27.13}$$

where $G^{(n)}$ is the updating gradient of the nth tensor, $Y_{(n)}$ is the original value of the nth tensor, and $Z_{(n)}$ is the recovery value of the nth tensor. The calculation method of $A^{(-n)}$ is as follows:

$$A^{(-n)} = A^{(N)} \odot \cdots \odot \ A^{(n+1)} \odot A^{(n-1)} \odot \cdots \odot A^{(1)}$$

$$(27.14)$$

Step5:   The $A^{(n)}$ is iteratively updated, and the program can be terminated in advance
when the target function meets the requirement.

27.3.3   Modified CPWOPT

The method of CPWOPT is first applied in the field of image restoration [4]. There
are still some problems when it is directly applied to the completion of traffic data
[15]. There is a temporal tendency in traffic data, which is not a random distribution.

Let $\mathcal{Y}$ be the weighted original velocity matrix. Given the tensor $\mathcal{Z}$ which is the
restored weighted velocity matrix of the same size as $\mathcal{Y}$. The computed loss function
based on the original CPWOPT method is as follows:

$$loss_{f_w} = -\frac{1}{2}\|\mathcal{Y} - \mathcal{Z}\|^2$$

$$(27.15)$$

To avoid using more time to randomly search for the optimized results, this paper
added four regular terms to separately describe the characteristics of weekday, time
period, speed difference and intrinsic speed class of road segment. The loss function
after adding a variety of the above regular terms is

$$loss\_f_w = -\frac{1}{2}\|\mathcal{Y} - \mathcal{Z}\|^2 + \frac{\lambda_1}{2}\|A - T_A\|^2 + \frac{\lambda_2}{2}\|B - T_B\|^2 + \frac{\lambda_3}{2}\|C - T_C\|^2 +$$

$$\frac{\lambda_4}{2}\|C - T_{class}\|^2 + p(\|A\|^2 + \|B\|^2 + \|C\|^2)$$

$$(27.16)$$

where T_A is the weekday factor matrix, which is calculated by the average speed of all
road segments per day; T_B is the time period characteristic matrix, which is calculated
by the average speed of particular time period of all road segments in all days; T_C is the
segment speed characteristic matrix, and the speed of each road segment is computed.
T_{class} is the intrinsic speed class characteristic matrix, expressed by the speed class
of the road segment; $\lambda_1, \lambda_2, \lambda_3$ and λ_4 are hyperparameters, indicating the weight of
the regular terms, p is the penalty coefficient, which is also a hyperparameter.

27.4 Experiments

The performance of the modified CPWOPT is evaluated in the traffic speed data of ride-hailing service collected by Didi Chuxing company. The data samples are GPS data of the ride-hailing vehicles collected of a week every 5 seconds, in the area around the Nanjing South railway station in June, 2018. Based on the geographic information provided by Didi company, the urban network is constructed by amount of connected links. The columns of the map matched data include the users' ID, link ID, collecting timestamp, point speed, point direction, line speed, status, last link id and so on. The users refer to the ride-hailing drivers of Didi company, and the data provided is preprocessed by the company and there is no information about the drivers. After detecting the links out of urban network, the target area has about 1537 links to be researched. Assuming the time period of 15 min, one day of 24 hours is divided into 96 time segments. By averaging the speed data of vehicles in a certain link in the certain time segment of a week, the tensor of size 7 * 96 * 1537 is obtained. The tensor needs to be recovered for the missing data of some links and some time periods.

To evaluate the performance of modified CPWOPT (MCPWOPT) in traffic speed data, the normal method of data completion and unmodified CPWOPT is applied in the same data sample to compare the results of data completion [16]. Most common data completion method is filling the missing data with mean value of the observed value. This paper not only compute the mean value, but also take the bias between different weekdays, different time segments and different links into consideration. The rank of CP decomposition is processed as a hyperparameter and auto-determined by the tool called Hyperopt in Python [17].

The root-mean-square error (RMSE) of recovered results is employed to compare three different methods [18]. The recovered results are more approaching true values when the value of RMSE is closes to zero. To evaluate the performance of the methods in different conditions, the missing ratio of the observed is manually set to be 20, 40, 60, and 80%. After utilizing the manually dropped data as test data, the computing method of the RMSE results is as follows [19]:

$$\text{RMSE} = \frac{\|\mathcal{Y}_m - \mathcal{Z}_m\|}{\|\mathcal{Y}_m\|}$$

$$(27.17)$$

where the $\mathcal{Y}_m$ is the tensor of manually dropped data, the $\mathcal{Z}_m$ is the recovered data at the same position of manually dropped data in tensor $\mathcal{Y}$, which is of the same size as $\mathcal{Y}_m$.

The results of RMSE of three methods are shown in Fig. 27.1. The commonly used method, filling missing data by mean value is not accurate compared with other methods. CPWOPT and MCPWOPT have stable performance when increasing the missing ratio from 20 to 80%. The MCPWOPT performs better when the ratio of

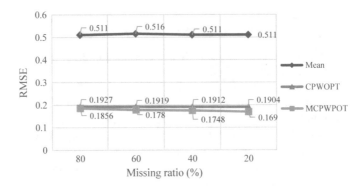

Fig. 27.1 RMSE for data completion

missing data decreases to 20%, and the performance will weaken when missing data starts increase.

27.5 Conclusions

This paper adopted the modified CPWOPT to recover the missing traffic speed data. After evaluating the performance of the model in the ride-hailing data collected by Didi, the results indicated that the method increase the accuracy of data completion compared to CPWOPT.

The research can be further improved on the two aspects. First, the value of the tensor rank can be increased gradually. Since the size of the original data tensor is 7 * 96 * 1537, the maximum value of rank is 7. After adding more traffic data, it is promised to increase the rank to improve the completion accuracy. Second, the CP decomposition combined with Bayesian model is desired to be experimented on the traffic speed data completion, which is demonstrated as an effective method in image restoration.

References

1. Kolda, T.G., Bader, B.W.: Tensor decompositions and applications. SIAM Rev. **51**(3), 455–500 (2009)
2. Gu, Z., Saberi, M., Sarvi, M., Liu, Z.: A big data approach for clustering and calibration of link fundamental diagrams for large-scale network simulation applications. Transp. Res. Procedia **23**, 901–921 (2017)
3. Tang, K., Chen, S., Liu, Z.: Citywide spatial-temporal travel time estimation using big and sparse trajectories. IEEE Trans. Intell. Transp. Syst. (99), 1–12 (2018)
4. Acar, E., Dunlavy, D.M., Kolda, T.G., Mørup, M.: Scalable tensor factorizations for incomplete data. Chemometr. Intell. Lab. Syst. **106**(1), 41–56 (2010)

5. Tucker, L.R.: Some mathematical notes on three-mode factor analysis. Psychometrika **31**(3), 279–311 (1966)
6. Bro, R.: PARAFAC. Tutorial and applications. Chemom. Intell. Lab. Syst. **38**(2), 149–171 (1997)
7. Liu, J., Musialski, P., Wonka, P., Ye, J.: Tensor completion for estimating missing values in visual data. IEEE Trans. Pattern Anal. Mach. Intell. **35**(1), 208–220 (2013)
8. Ran, B., Tan, H., Feng, J., Liu, Y., Wang, W.: Traffic speed data imputation method based on tensor completion. Comput. Intell. Neurosci. **2015**(2), 22 (2015)
9. Wang, Y., Zheng, Y., Xue, Y.: Travel time estimation of a path using sparse trajectories. In: Proceedings of the 20th ACM SIGKDD International Conference on Knowledge Discovery and Data Mining (2014)
10. Tang, K., Chen, S., Liu, Z., Khattak, A.J.: A tensor-based Bayesian probabilistic model for citywide personalized travel time estimation. Transp. Res. Part C Emerg. Technol. **90**, 260–280 (2018)
11. Zhao, Q., Zhang, L., Cichocki, A.: Bayesian CP factorization of incomplete tensors with automatic rank determination. IEEE Trans. Pattern Anal. Mach. Intell. **37**(9), 1751–1763 (2015)
12. Wang, C., Xu, C., Xia, J., Qian, Z., Lu, L.: A combined use of microscopic traffic simulation and extreme value methods for traffic safety evaluation. Transp. Res. Part C Emerg. Technol. **90**, 281–291 (2018)
13. Zhang, J., Qu, X., Wang, S.: Reproducible generation of experimental data sample for calibrating traffic flow fundamental diagram. Transp. Res. Part A Policy Pract. **111**, 41–52 (2018)
14. Liu, Y., Jia, R., Xie, X., Liu, Z.: A two-stage destination prediction framework of shared bicycles based on geographical position recommendation. IEEE Intell. Transp. Syst. Mag. **11**(1), 42–47 (2019)
15. Wang, C., Xu, C., Dai, Y.: A crash prediction method based on bivariate extreme value theory and video-based vehicle trajectory data. Accid. Anal. Prev. **123**, 365–373 (2019)
16. Liu, Z., Chen, X., Meng, Q., Kim, I.: Remote park-and-ride network equilibrium model and its applications. Transp. Res. Part B Methodol. **117**, 37–62 (2018)
17. Zhou, M., Qu, X., Li, X.: A recurrent neural network based microscopic car following model to predict traffic oscillation. Transp. Res. Part C Emerg. Technol. **84**, 245–264 (2017)
18. Qu, X., Wang, S., Zhang, J.: On the fundamental diagram for freeway traffic: a novel calibration approach for single-regime models. Transp. Res. Part B Methodol. **73**, 91–102 (2015)
19. Liu, Z., Wang, S., Chen, W., Zheng, Y.: Willingness to board: a novel concept for modeling queuing up passengers. Transp. Res. Part B Methodol. **90**, 70–82 (2016)

Author Index

© Springer Nature Singapore Pte Ltd. 2019
X. Qu et al. (eds.), *Smart Transportation Systems 2019*, Smart Innovation,
Systems and Technologies 149, https://doi.org/10.1007/978-981-13-8683-1

Printed in the United States
By Bookmasters